THE LONGEST EXPERIMENT

For those who began the measurement,
and those who kept measuring.

For everyone who continued
after the reason for beginning had changed.

For those who leave behind work
they will never see completed.

For those who kept looking.

THE LONGEST EXPERIMENT

Antonio Pascarella

StoneGate Publishing, LLC – Meriden, Connecticut

The Longest Experiment
By Antonio Pascarella

This book is a work of narrative nonfiction. It draws on publicly available scientific research and institutional sources. Any interpretations, syntheses, or framing of that material are the author's own.

The information presented herein is intended for educational and informational purposes only. It does not constitute scientific, legal, financial, or policy advice.

While every effort has been made to ensure accuracy, the author and publisher make no representations or warranties with respect to the completeness or suitability of the information contained in this book.

Published by StoneGate Publishing, LLC
www.stonegatepublishing.com
Permissions & inquiries:
permissions@stonegatepublishing.com

Cover image: Lunar photograph by Antonio Pascarella, ©2007. Used by permission.

ISBNs:
Hardcover: 979-8-9999855-9-0

Library of Congress Control Number: 2026934171

First edition, March 2026 – Printed in the United States of America

Printer's key: 10 9 8 7 6 5 4 3 2 1

PHASE 1 — THE MEASUREMENTS

THE DISTANCE

They fired the laser every night, whether anyone was watching or not.

The observatory sat on a ridge above San José, far enough from the city that the lights softened instead of vanished. From the outside it looked like a relic—white dome, narrow slit, a shape better suited to the twentieth century than the twenty-first. Inside, the work was quieter still. There were no crowds, no countdowns, no sense of ceremony. Just schedules. Just repetition.

At 12:03 a.m., the system aligned.

Calibration flags cleared one by one.

Atmospheric variance fell within tolerance.

Timing gates locked, synchronized to the master clock.

Nothing else changed.

A pulse of green light rose through the slit in the dome, cut cleanly through the atmosphere, and vanished into black. It traveled farther than any human had ever walked, farther than any aircraft could fly, farther than any satellite could linger without falling.

A little more than a second later, it struck the Moon.

Not the Moon as it appeared to the naked eye—bright, symbolic, familiar—but a precise location in the Sea of Tranquility, where a flat aluminum panel no larger than a suitcase sat bolted into ancient dust. The panel did nothing. It absorbed no energy. It made no decisions. It simply returned the light exactly the way it came.

Two and a half seconds after it left Earth, the pulse came back.

Inside the control room, the return registered as a soft spike on a monitor—an upward flicker that barely disturbed the surrounding data noise. There was no visible burst of light and nothing to mark the moment as an event, nothing to suggest that anything at all had happened. No sound followed. No movement. Just a timestamp, a number, and a single new line appended to a dataset that had been growing, methodically and without interruption, since the summer of 1969.

“Return confirmed,” someone said, already looking away.

The technician at the console logged the result without comment. He had been doing this long enough that the motion required no thought. Confirm. Save. Advance. The system would repeat the process automatically, cycling through corrections for atmospheric distortion, Earth’s rotation, orbital eccentricity. The raw number would be cleaned, averaged, folded into a model.

Error margins were recalculated and flagged for review.
Residual noise was stripped away by routines older than anyone in the room.
The measurement took its place among thousands of others, indistinguishable at a glance.
No alarms triggered.

Nothing about it was exciting.

The Moon had not changed its face. The night sky looked exactly as it had the night before, and the night before that. Outside the dome, the city continued its low, constant hum. Cars moved. Servers processed transactions. People slept.

On the screen, the distance updated.

It was a few millimeters greater than the same measurement taken the year before.

As expected.

The technician paused, just long enough to notice the pause. It wasn’t concern that stopped him—it was the peculiar discomfort of permanence. The kind that came from systems that worked too well, from processes that ran so smoothly they stopped feeling like work and started feeling like background.

“Still clean?” he asked.

“Still clean,” came the reply.

No anomalies.
No alerts.
No reason to escalate.

The log remained unmarked beyond its standard entry.
Screens dimmed slightly as the cycle completed.

Chairs creaked as weight shifted, then settled again.
The room returned to its steady watch.

The mirrors never failed. They had no power source to degrade, no circuitry to age. They didn't drift or tilt or erode. They sat exactly where they had been placed, reflecting light back to its origin with mathematical indifference.

Someone once called it humanity's longest-running experiment. That wasn't quite right. The experiment had ended decades ago, its success already established.

What remained was the measurement.

Every night, light left Earth.
Every night, light returned.
And every night, the time it took told the same quiet story.

A graph updated automatically in a system no one in the room had built and no one fully owned anymore. The line on it rose at an angle so slight it looked flat unless you zoomed in, unless you already knew the scale, unless you were looking for movement where intuition insisted there should be none.

The Moon was moving away.

Not suddenly.
Not dangerously.
Not in any way that disturbed the sky.

It always had been.

What had changed wasn't the motion, but the certainty. For the first time in human history, the distance between Earth and its only natural satellite was no longer inferred from equations or celestial mechanics alone. It was measured directly, with light and time, again and again, without interruption.

It was recorded.
Archived.
Cross-checked.

And because it behaved exactly as predicted, it was treated as settled.

The technician closed the file and moved on to the next task. Tomorrow the number would be averaged. Next month it would be cited.

Next year it would appear in a paper that confirmed what everyone already knew.

It wouldn't make the news.

It never had.

Outside, the Moon hung where it always had, bright enough to cast shadows, unchanged to the eye, carrying with it centuries of stories, calendars, and assumptions. No one looking up would have guessed that anything about it was different.

The sky offered no warning.

Only the instruments did.

And they spoke in millimeters.

∝ ∝ ∝

THE MIRRORS

They were not designed to be impressive.

That was the first thing people misunderstood later, when the footage was replayed and compressed into anniversary specials and grainy clips shared online. Viewers expected something monumental, some visible proof of triumph carefully framed for the camera, evident symbol of human ambition. What they saw instead looked like equipment. Flat. Dull. Practical.

The mirrors were carried to the surface of the Moon like luggage.

They were unpacked with the same care, handled deliberately. Nothing about the task suggested it would have been remembered.

Neil Armstrong remembered thinking they looked fragile, though he knew they weren't. One hundred corner-cube prisms mounted in a simple aluminum frame, each one polished to a precision that bordered on absurd. The tolerances were tighter than anything he'd handled in flight training. Geometry so exact it felt almost arrogant, as if it assumed the universe would behave.

"Let's get this level," Aldrin said, his voice calm, clipped, professional.

They worked slowly. Not ceremonially—there was no time for that—but carefully, the way people do when they know a mistake will be permanent. The lunar surface resisted them, powdery and uneven, a fine gray dust that clung to boots and gloves and refused to settle. Armstrong adjusted the frame by millimeters, checking alignment against Earth's predicted position, not where it was now but where it would be by the time the light arrived.

The mirrors didn't point at Houston.

They pointed at motion.

No one watching at home understood what was happening. The broadcast cut away quickly, commentators filling the space with reassurances and wonder. There was a flag to plant. Words to say. History to make.

This was just instrumentation.

When the array was finally secured, Aldrin straightened and dusted his gloves together, a small, habitual gesture that meant nothing and everything. Armstrong gave the frame one last look—not pride, not sentiment, just confirmation.

"Array deployed," Aldrin said.

In Houston, a technician checked a box.

And the world moved on.

∝ ∝ ∝

The first successful return came weeks later, not from Texas or Florida but from California, where the air was thinner and the sky cooperated more often, on average. The choice had been pragmatic rather than symbolic. Fewer clouds. More stable conditions. Less atmosphere to interfere with a signal that would already be vanishingly faint by the time it returned.

The laser itself was unimpressive. A pulse so brief and narrow it could disappear into the night without anyone noticing, swallowed immediately by distance and darkness. There was no beam to watch, no visible streak across the sky. To anyone standing outside the observatory, nothing appeared to happen at all.

Inside, the sequence unfolded exactly as planned.

The pulse left the instrument, crossed the void, struck the mirrors, and came back diminished but intact. The return was weak—far weaker than the outgoing signal—but it was clean. Recognizable. Distinct from noise. It arrived not as a glow or a flash, but as a timing event, a precise registration against an internal clock that had been waiting for it.

Time did the rest.

They measured the delay not with excitement but with disbelief. The initial readout was checked, then checked again. Parameters were verified. Calibration constants revisited. No one wanted to be the first to say it aloud. The result was too coherent, too well-behaved, too exact to trust immediately.

The error bars collapsed in on themselves like a bad argument. Where there had once been estimates, there was now a number. Where there had been theory, there was now repetition. The signal didn't merely return once; it returned again, and again, with timing that refused to drift.

They fired again.

And again.

The distance held.

At first, the agreement was within centimeters—already better than anyone had reasonably expected. Then millimeters. Then less. The uncertainty shrank because the measurements left it no room to survive. Each return narrowed the space in which doubt could exist.

There was no dramatic moment of realization. No applause. No raised voices. Just a gradual, collective understanding that something irreversible had happened.

It was immediately clear what this meant. Astronomy had crossed a line, no longer content to describe the universe from a distance, no longer limited to inference and approximation, but beginning to anchor its claims to something firmer. The Moon was no longer an object you reconstructed through math and motion alone. It was something you could touch with light and hear respond.

Not metaphorically. Practically.

The delay could be measured. Argued over. Measured again. The round-trip time—there and back—collapsed an astronomical distance into a number small enough to fit on a printout, small enough to be compared, archived, and revisited. A distance that had once belonged to diagrams and imagination now lived inside a laboratory instrument.

They ran controls. They swapped components. They adjusted for atmospheric effects, for Earth's rotation, for orbital motion. The number shifted slightly, predictably, in ways the equations had already prepared them for. Nothing about the result resisted explanation.

That was what made it unsettling.

The measurement didn't challenge existing physics. It confirmed it. And in doing so, it changed the relationship between humanity and the space it occupied. The Moon was no longer simply *there*, moving according to celestial rules. It was *here*, participating in an exchange, responding on command, answering a question with a delay that could be timed to fractions of a microsecond.

A ruler had been placed between worlds.

Not a metaphorical one. A literal one, calibrated in light and time, extending across vacuum without ever leaving the ground. Distance had been domesticated, not eliminated, but rendered legible in a way it had never been before.

They documented the results carefully. Conservatively. The language was restrained. Claims were framed narrowly, just as they had been taught. But beneath the formality, the implication remained unchanged.

Something that had always been immeasurable in any practical sense had just been measured.

And it would never return to being abstract again.

Years passed. Missions ended. The last men left the lunar surface and didn't return. The sense of proximity that had once made the Moon feel reachable receded with them, replaced by distance not just of space, but of attention. Political priorities shifted. Budgets reallocated. Agencies reorganized themselves around new threats, new ambitions, new timelines. The Moon remained, pale and unchanged, but no longer urgent.

The mirrors stayed.

They required no maintenance. No upgrades. No calibration schedules. No review panels. They sat in vacuum and silence, untouched by weather, immune to erosion and decay, indifferent to radiation and micrometeoroids in ways few engineered objects could claim. There were no moving parts to seize, no joints to loosen, no power systems to degrade. They reflected light because geometry demanded it.

While instruments on Earth were replaced and modernized and renamed—each generation more precise, more complex, more fragile—the mirrors remained exactly what they had been on the day they were set down. They didn't benefit from technological progress, but neither were they threatened by it. They existed outside cycles of obsolescence.

Simple. Passive. Exact.

Their simplicity became a kind of invisibility.

The people who knew about them in any meaningful sense grew older. The engineers who had designed them, the machinists who had polished their faces to tolerances measured in wavelengths, the astronauts who had carried them across regolith and aligned them by hand—one by one, they retired. Their names receded from authorship lists. Their offices were reassigned. Their institutional memory was reduced to archived memos and citations stripped of context.

Those who followed encountered the mirrors indirectly. As diagrams in textbooks. As footnotes in review articles—quickly passed, rarely lingered over. As references embedded so deeply in methodology sections that they felt less like experiments than assumptions. The mirrors were no longer something one *worked on*. They were something one *worked with*.

Laser ranging became routine.

What had once required specialized teams and dedicated nights at observatories became automated, scheduled, expected. The pulses were fired. The returns recorded. The numbers logged. The process ran quietly in the background of research agendas focused elsewhere. Studies of gravity. Of relativity. Of Earth's interior. Of orbital mechanics refined to decimal places few outside the field could interpret.

The mirrors did their part without variation.

The Moon's distance was recorded alongside dozens of other variables—tectonic motion, atmospheric loading, rotational wobble—one more column in a dataset too large for any one person to hold in their head. The datasets grew. Software improved. Visualization tools became more powerful, capable of rendering patterns across timescales that once required years of manual work.

Yet familiarity dulled attention. What had been astonishing became expected. What was expected ceased to provoke curiosity.

The experiment persisted, not as an object of inquiry, but as infrastructure.

Its results were trusted because they had always been trustworthy. Its methods were unquestioned because they had always worked. The mirrors returned exactly what they were asked for, no more and no less, each time. The process asked nothing of interpretation, only consistency. In that reliability, nothing seemed to require attention. There were no anomalies dramatic enough to demand intervention, no failures sharp enough to require explanation.

Their perfection insulated them from scrutiny.

Time passed in quieter ways.

Graduate students learned the technique without learning its history. Postdoctoral researchers used the data without ever asking how long it had been accumulating. Senior scientists referenced trends spanning decades without dwelling on what it meant for an experiment to run longer than a human career, longer than a research program, longer than most institutions were designed to remember.

The mirrors waited.

They had no sense of accumulation. No awareness of duration. Each return was identical to the last—an isolated event, complete unto itself, carrying no memory of what had come before, beyond the moment of measurement itself, alone. The mirrors didn't know that their answers stretched across decades, or that repetition itself was becoming a form of evidence. They didn't know that their returns now spanned generations, or that the line they traced through time had become long enough to invite interpretations that couldn't be resolved within a single grant cycle.

They reflected light.

That was all they had ever been asked to do.

And because they asked for nothing in return—no maintenance, no justification, no attention—they received exactly what they required: neglect disguised as trust.

Decades later, a graduate student in a windowless lab would pause over a plot that showed the accumulated measurements across half a century. The lab was lit by overhead panels calibrated to reduce glare, the walls lined with equipment racks and whiteboards filled with equations that had been written and erased too many times to feel provisional, anymore. There was no sense of history in the room. Everything about it was designed for continuity, not reflection. The displays refreshed on fixed intervals, numbers updating without emphasis.

Nothing in the workflow marked what they represented.

She wasn't looking for anomalies. She was cleaning data, preparing it for a paper that would cite older papers that cited even older ones. The task was procedural, almost mechanical. Remove outliers. Normalize timestamps. Confirm that the metadata matched the assumptions built into the model. It was the kind of work that rewarded consistency rather than insight.

Her finger traced the line on the screen without thinking.

"It's still moving," she said, mostly to herself.

Her advisor didn't look up from his monitor. "Of course it is."

"I know," she said. "I just mean… it always has."

The advisor nodded, distracted, already half inside another problem. "Tidal transfer. Angular momentum. First-year stuff."

She didn't argue. There was nothing to argue about. The physics was clean. The explanation elegant. Earth slowed, and the Moon drifted outward. Energy moved. It moved where the equations said it should move. A closed system doing exactly what it was supposed to do.

The line on the screen reflected that confidence. Over decades, it rose smoothly, obediently. It didn't spike. It didn't wobble. It didn't challenge the framework that contained it.

Still, she zoomed in.

The axes recalibrated, the noise floor dropping away. Error bars tightened, revealing structure the overview concealed. What had looked settled resolved into something less certain.

At full scale, the line was flat. At higher resolution, it wasn't.

The difference was small enough to be uninteresting. The sort of deviation that vanished when averaged, that disappeared into uncertainty bars when presented properly. She knew how to make it behave. She had been trained to do exactly that.

She adjusted the axes. The curve smoothed. She adjusted them again. It sharpened slightly, then flattened once more. The movement never disappeared entirely, but it could be made negligible with the right framing.

She told herself this was expected. The models accounted for it. The literature described it. Nothing here contradicted anything she had been taught.

She zoomed back out.

The line returned to its reassuring shape, stretching calmly across the screen, its slope gentle enough to feel almost static. The software overlaid trend markers automatically. Confidence intervals settled into place. Everything aligned, not as an achievement, but as confirmation.

She saved the image and moved on.

The next dataset required the same treatment. The same cleaning. The same quiet confidence that nothing unexpected would appear. She worked methodically, letting familiarity guide her hands. There was no reason to linger. The schedule was tight. The submission deadline closer than it felt.

Occasionally, her eyes flicked back to the earlier plot. Not with concern. With recognition. The kind that comes from seeing something too often to be surprised by it.

The advisor eventually glanced over. "Everything good?"

"Yeah," she said. "Looks fine."

He nodded, satisfied. "Good. We'll fold it in with the rest."

Fold it in. The phrase carried no weight. It meant alignment, consistency, conformity. It meant the data would join the others without friction, indistinguishable once processed, one more contribution to a body of work already too large to feel personal.

She watched as the file joined a folder of similar plots, each named according to a convention that stripped away context in favor of order. Dates. Identifiers. Version numbers. The system didn't care which dataset had caught her eye, or why.

She returned to her work.

The lab hummed quietly. Cooling fans cycled. Screens refreshed. Somewhere else in the building, another student ran a simulation that assumed the same constants, the same rates, the same long view.

She thought briefly about how long the measurements had been running. Not in detail. Just long enough to feel abstract. Longer than her program. Longer than her advisor's early career. Longer than some of the software she was using had existed.

The thought didn't settle. It didn't demand action.

It registered and passed, like a variable noted and set aside. The workflow resumed without interruption.

She finished the task, closed the file, and logged the update according to protocol. The system accepted it without comment.

When she shut down her workstation at the end of the day, the plots vanished from the screen. The line persisted elsewhere, stored redundantly, backed up automatically, waiting to be called up again when needed.

She didn't think about it on the walk home.

She didn't need to.

The experiment continued.

∝ ∝ ∝

The mirrors never warned anyone.

They never signaled danger or urgency or loss. They never deviated from expectation, never behaved in a way that suggested consequence. Each time a pulse reached them, they did exactly what they had been designed to do, receiving it and returning it unchanged. They didn't amplify the signal, distort it, or comment on it in any way that could be interpreted as concern or confirmation.

They reflected light and nothing else.

Their silence wasn't concealment or refusal. It wasn't restraint or discretion. It was simply the absence of intent. The mirrors didn't choose what to return. They didn't distinguish between one measurement and the next within their sequence. Geometry governed their response, and geometry alone. Whatever reached them at the proper angle came back according to the same rules every time.

Whatever meaning emerged from the measurements had to be supplied by the people watching the numbers accumulate.

At first, that work belonged to a small group of specialists who knew the equipment intimately. They understood the geometry of the stations, the timing offsets between clocks separated by continents, the atmospheric corrections required to account for air that bent light differently on different nights. They knew how fragile the precision was, how many small adjustments had to align before a clean return could be fully trusted.

Each data point carried with it an awareness of effort, of limitation, of how easily the result could have failed.

In those early years, the measurements were examined closely because they were new. Each successful return felt provisional, something that might still collapse under scrutiny. The numbers were checked against expectations, against independent calculations, against one another. Confidence grew slowly, reinforced by repetition rather than surprise.

Later, the work changed hands. Responsibility diffused. Tasks were subdivided, documented, reassigned. What mattered was that the sequence held.

Understanding became optional as long as results remained consistent.

What had once required intimate familiarity with the apparatus became institutional knowledge, embedded in procedures and documentation. The reasoning that had guided early interpretation was compressed into protocols, then inherited by analysts who trusted the process without ever having met its architects. Training focused on execution rather than origin. The system was presented as complete.

Over time, the measurements lost their novelty.

Laser ranging became routine, folded into broader studies of gravity and relativity and Earth's interior. It became one variable among many in increasingly dense datasets, one input feeding models whose primary questions lay elsewhere. The Moon's distance appeared as a column in a table, updated automatically, reviewed occasionally, rarely questioned. It was there because it had always been there.

The system worked. That alone was enough.

Oversight narrowed to maintenance of cadence.

Outputs were checked for continuity, not meaning.

Attention followed novelty elsewhere.

Familiar signals blended into background expectation.

What persisted did so without argument.

Reliability required no audience.

Years passed. Careers ended. Funding cycles came and went. Research priorities shifted in response to new discoveries, new instruments, new questions that promised quicker returns. New generations of scientists learned about the experiment the way one learns about an old bridge or a buried cable—important, foundational, but finished. Its continued operation was treated as evidence of its completeness.

If the numbers drifted, the assumption was calibration. If the trend persisted, the explanation was theory. Deviations were absorbed into models that already expected motion, change, adjustment. No one expected revelation from a tool designed to confirm what was already believed.

The mirrors didn't question anything.

They didn't distinguish between expectations and anomalies, between confirmations and implications. They didn't know what was being tested, only that a signal had arrived and it required an answer. That answer would be governed by fixed geometry, calibrated clocks, and atmospheric correction tables. Whether the result reinforced certainty or quietly complicated it wasn't part of their function.

They responded with the same precision regardless of context. A pulse sent as part of a major study was returned no differently than one fired during a routine check. The mirrors didn't care whether the data would be examined closely, summarized briefly, or passed downstream into a system that would flatten it into averages and trends.

They didn't know if the question was trivial or profound, or merely a noise that demanded attention. They didn't know whether the answer would be ignored, misunderstood, or deferred. They offered no signal that interpretation should slow down or stop. No cue that the meaning of what was being measured might extend beyond the framework that had first justified the measurement.

They simply answered.

And because they did, consistently and without interruption, responsibility for meaning shifted to the observers—to their habits, assumptions, and systems they trusted to decide what deserved attention.

∝ ∝ ∝

The Line That Wouldn't Sit Still

The problem with long-running experiments wasn't that they failed.

It was that they became invisible.

Familiarity dulled attention more effectively than failure ever could. What endured without incident faded into assumption.

By the time the meeting was convened, the dataset had outlived the people who started it. The original engineers were gone, their notebooks archived, their assumptions embedded so deeply into software and process that no one thought to question them. The laser-ranging program persisted because it worked, because it was cheap, because it required no imagination to maintain.

The conference room smelled faintly of coffee and old carpet, unchanged by the updates that were meant to modernize it. Meant for time to feel procedural, not urgent. Someone had tried to modernize it with a glass table and thin metal chairs, but the walls were still thick, the ceiling still low. There were no windows. There never were.

On the screen at the front of the room, the graph waited.

It had been rendered a thousand times before, but rarely like this. Someone had taken the trouble to strip it down—no annotations, no color coding, no explanatory labels beyond the bare minimum. Time on the horizontal axis. Distance on the vertical. A single line, drawn carefully enough that no one could accuse it of exaggeration.

At first glance, it looked flat.

"That's the full span?" someone asked.

"Yes," said Martin Keller, who had been standing near the screen longer than was comfortable. He hadn't planned to be the one presenting. He rarely volunteered for it. His job was to make sure the numbers behaved before anyone else saw them.

He tapped the remote. The scale changed.

The line tilted upward.

Not dramatically. Not enough to provoke gasps or alarm. Just enough that it could no longer be dismissed as an artifact of resolution.

"That's sixty-two years," Keller said. "Continuous measurement. Same targets. Same method. Different instruments, obviously, but cross-calibrated."

A woman near the end of the table frowned. "We know the Moon is receding."

"Yes," Keller said. "Everyone in this room knows that."

No one contradicted him.

The physics had been settled long before any of them were born. Tidal friction transferred angular momentum from Earth's rotation to the Moon's orbit. Earth slowed. The Moon drifted outward. Elegant. Closed. Predictable.

First-year stuff.

"What's the issue, then?" someone asked.

Keller hesitated. He didn't know the answer, but saying it out loud changed the nature of the meeting. He had learned, over time, that some conversations were safer when they stayed abstract.

"The issue," he said finally, "is that the line doesn't sit still when you ask different questions of it."

He changed the slide.

Now there were several lines, nearly overlapping. Slight variations in processing. Different correction models. Independent analyses run by different teams.

They all rose.

The separation was visible when the scale tightened. Differences that once averaged out now refused to disappear.

"The rate is consistent," Keller continued. "That's not the concern. What's changed is our ability to pretend this is just background."

Someone leaned back in their chair. "You're talking about sensitivity again."

"I'm talking about assumptions," Keller said.

He pulled up another graph, this one messier, less intuitive. It showed modeled Earth systems—axial stability, precession, seasonal variance—plotted not against time but against perturbation.

"The Moon doesn't just affect tides," he said. "It damps chaos. It smooths out wobble. It keeps Earth's tilt in a narrow range."

"That's well understood," the woman said.

"It's understood in theory," Keller replied. "What we haven't done is stress-test what happens when that stabilizing influence weakens incrementally but persistently."

A man in a dark suit cleared his throat. He hadn't spoken yet. He rarely did. His role was to listen until something crossed a line.

"You're not suggesting anything is imminent," he said.

"No," Keller said quickly. "No one is."

The room relaxed, just a little.

"What I'm saying," Keller continued, more carefully now, "is that our long-horizon models assume a level of orbital stability that is historically justified but not guaranteed. The Moon's recession is slow, yes. But the systems it stabilizes are non-linear."

Silence.

Someone scribbled something in a notebook they would never look at again.

"So what?" the man in the suit asked.

Keller met his eyes. "So the error bars widen. Not tomorrow. Not next year. But they widen."

"And when do they matter?"

Keller didn't answer right away. He walked back to the screen and zoomed in, not on the future, but on the present—on the most recent decade of data.

"Here," he said. "Right here is where the line starts refusing to behave like trivia."

The movement was still small. Millimeters per year. Numbers no one outside the room would lose sleep over.

But the line no longer looked like noise.

It looked like motion.

∝ ∝ ∝

After the meeting, Keller lingered in the hallway, watching people disperse in the practiced way of institutions that had learned to absorb discomfort. Doors opened and closed with muted clicks. Voices softened as conversations fractured into pairs. No sense of rupture, no visible sign that anything meaningful had occurred. The meeting had ended cleanly, on schedule, with notes taken and next steps outlined.

No one argued. No one dismissed him outright.

That was almost worse.

Disagreement, at least, would have implied engagement. Dismissal would have clarified boundaries. What he had encountered instead was accommodation—the kind that allowed ideas to enter the room, circulate briefly, and then settle into procedural sediment where they could be examined later, indefinitely.

A colleague fell into step beside him as they moved toward the elevators. Her pace matched his without effort, as though the moment had been anticipated.

"You pushed that pretty far," she said.

"I didn't push it at all," Keller replied. "I just showed them what's already there."

She gave a small nod, neither agreeing nor contradicting him. "You know how this goes. They'll ask for more review. More modeling. Another panel."

"I know."

"And you'll give it to them."

"Yes."

The answer came without resentment. This wasn't a battle he was unfamiliar with. He had spent years translating uncertainty into acceptable language, years learning how to present results in ways that invited continuation rather than resistance. Requests for additional work were not obstacles; they were the currency of credibility.

They walked in silence for a few steps. The hallway lights hummed faintly overhead, set to an intensity designed to be noticed by no one.

She hesitated, then spoke again. "Do you think this is something we should be… louder about?"

Keller stopped walking.

The pause was subtle, but deliberate. For a moment, the hallway felt narrower than it had before, its clean lines closing in around a question that had not been raised during the meeting itself. It didn't fit neatly into an agenda item.

"I think," he said slowly, choosing each word with care, "that the problem isn't that we don't know enough. It's that we know just enough to be uncomfortable, and not enough to be certain."

He could hear how careful it sounded, even as he said it. How reasonable. How easily it could be repeated without committing anyone to action.

She considered that. "That's not usually when institutions act."

"No," Keller agreed. "That's when they wait."

They stood there for a moment longer, neither of them in a hurry to move on. Around them, the building resumed its rhythms. Phones vibrated. Elevators chimed. Somewhere, another meeting began on another floor, its participants unaware that anything unusual had been discussed just minutes earlier.

Eventually, she nodded once more. "Well. Let me know what they come back with."

"I will."

She turned and headed toward the stairs, her footsteps receding quickly, absorbed into the ordinary noise of the building. Keller watched her go, then resumed walking himself.

As he moved, he replayed the meeting in fragments. The questions had been polite. The concerns well framed. No one had challenged the data. No one had questioned the methodology. The hesitation had not been scientific. It had been procedural.

More modeling. More review.

Each request would be justified. Each would improve resolution, tighten confidence intervals, reduce ambiguity. None would resolve the underlying discomfort that had settled over the room when the implications had been allowed to surface, briefly, without conclusion.

Keller reached the end of the hallway and paused again, this time at the window overlooking the parking structure below. The view was unremarkable. Concrete. Lines. Order. A system designed to manage flow without calling attention to itself.

He understood why waiting felt appropriate. Uncertainty demanded caution. Caution demanded process. Process demanded time. None of this violated institutional logic. In fact, it reinforced it.

The difficulty was that time was already part of the measurement.

He exhaled slowly, then turned away from the window. There would be follow-up emails. Requests for clarification. Invitations to participate in working groups that would meet quarterly and generate documents that summarized progress without altering direction.

He would attend them. He would answer questions. He would provide what was asked for.

That was how this worked.

The institution would wait because waiting preserved coherence. It allowed knowledge to accumulate without forcing a reckoning. It kept uncertainty contained within acceptable bounds.

And for now, that was enough.

∝ ∝ ∝

That night, back in his office, Keller pulled up the dataset again. Not the cleaned version. The raw numbers. The timestamps. The returns.

Columns stacked endlessly downward, each row another pulse of light sent outward and received back, logged without interpretation. No smoothing. No averaging. Just time, distance, and the quiet arithmetic that connected them.

Light leaving Earth.
Light coming back.
Time accumulating.

The early entries were coarse, their margins wide with uncertainty. Later ones narrowed, sharpened. The instruments had improved. The clocks had grown more precise. The error bars had shrunk until the motion emerged not as a suggestion, but as a slope—gentle, persistent, impossible to explain away.

He scrolled slowly, watching years pass in seconds. Administrations changed. Funding cycles rose and fell. Instruments were upgraded, recalibrated, replaced. The mirrors remained.

He thought about them now—silent, indifferent, still doing exactly what they had been designed to do. They didn't care how the data was framed. They didn't care who looked at it or who didn't. They didn't know they were being cited, or debated, or quietly set aside in favor of more urgent work.

They just reflected.

Keller leaned back in his chair and stared at the screen until the line blurred slightly, not from motion but from fatigue. The office was dark except for the monitor glow and the faint spill of light from the hallway beyond his door. Somewhere in the building, a ventilation unit cycled on and off, steady as a metronome.

The Moon, unchanged to the eye, continued its slow retreat, indifferent to memos and human scales of urgency. It didn't hurry. It didn't pause for human attention. It obeyed only the long, patient rules that had been written before anyone thought to measure them this precisely.

He had seen the argument before, in conference rooms and review panels. The numbers were real, everyone agreed. The explanation was the problem. Tides, thermal effects, instrumentation drift. Each accounted for a portion. None closed the gap entirely. The remainder was always small enough to defer, to footnote, to assign to future work.

Future work had arrived.

Keller zoomed the plot until individual years dissolved into continuity. The line didn't waver. It didn't spike. It didn't announce itself. It simply kept going, a quiet record of distance increasing one fraction of a centimeter at a time.

The line would keep moving.

He wondered how many people had looked at this same curve over the decades and decided it was someone else's problem. How many had meant to come back to it later, when there was more time, better tools, a clearer theoretical frame.

Later had come.

The question wasn't whether it would matter.

It was when people would stop treating that question as theoretical.

α α α

Margin of Error

The mistake people made, later, was assuming precision brought comfort.

It didn't.

Precision brought responsibility.

The lab where the data was processed had once been a storage space. The ceiling was low, the lighting practical, the walls painted a color meant to discourage lingering. It worked. People came in, did their work, and left.

The machines stayed. They ran continuously, indifferent to who operated them. Logs accumulated without commentary. Corrections were applied, documented, and forgotten. Nothing in the room suggested that the numbers required anything more.

Rows of monitors glowed softly, each one showing something slightly different: raw returns, corrected signals, statistical summaries. The numbers were small. Ridiculously small. Distances measured in meters, shifts measured in millimeters, uncertainties measured in fractions of both.

"This is why no one panics," said Elena Ortiz, tapping the screen with the back of her pen. "Because this looks like nothing."

She had been with the program long enough to remember when the error bars were larger than the signal. Back then, disagreement had been easy. You could argue methodology, instrumentation, atmospheric noise. You could say *we don't know yet* and mean it.

Now, the margins had collapsed. Variance that once masked disagreement had been stripped away.
What remained was precision without buffer.

The laser pulses returned with clockwork reliability. Atmospheric corrections were better understood. Earth's rotation, wobble, tidal deformation—everything that could be modeled had been, refined again and again by people who never expected their work to leave the footnotes.

"What's the current uncertainty?" she asked.

"Sub-millimeter," replied Jonah Park without looking up. "On the annual mean."

Elena exhaled through her nose. "That's absurd."

"Accurate," Jonah said. "But yes. Absurd."

That was the problem. When your uncertainty dropped below intuition, people stopped trusting their instincts and started trusting the numbers instead. And the numbers, infuriatingly, refused to sit still.

Elena pulled up the long view—not the graph everyone liked, flattened and reassuring, but one that made sense if you believed the data mattered.

Sixty years of returns. Each one timestamped. Each one corrected. Each one archived.

A line, rising gently.

"The Moon moves away," she said, not as a revelation but as a reminder. "That's not new."

Jonah nodded. "What's new is how little room there is left to argue about it."

When the mirrors were first installed, no one expected this level of clarity. The experiment had been ambitious, yes, but even its designers assumed the data would be noisy, that lunar dust or thermal stress or some unforeseen variable would complicate things. Instead, the opposite happened.

The universe cooperated.

Light behaved.

Time behaved.

The mirrors behaved.

Every improvement in instrumentation narrowed the space where doubt could live. Where once you could hide behind uncertainty, now there was only interpretation.

Interpretation was where things got uncomfortable.

"Error bars used to protect us," Elena said later, during a meeting that had drifted off its original agenda. "They gave us room to be wrong."

"And now?" asked someone from the back of the room.

"Now they box us in."

She gestured to the screen. "When the margin of error is larger than the effect, you can argue about whether the effect is real. When the margin collapses below it, the effect becomes a fact. And facts don't care whether you're ready for them."

No one disagreed. That wasn't the same thing as agreement.

∝ ∝ ∝

Outside the lab, the world went on, largely indifferent to the differences between precision and certainty. Construction crews poured concrete according to codes and tolerances that assumed the ground stable, that treated subsidence and drift as local problems with known bounds. Blueprints specified margins that had been tested repeatedly, reinforced by decades of compliance and revision. Buildings rose and held, reinforcing the belief that the assumptions beneath them were sound.

Farmers planted on schedules inherited from parents and grandparents, trusting seasons that had arrived on time for generations. The calendars they followed were not abstractions but habits, refined by experience and reinforced by success. Planting too early or too late carried consequences immediate enough to discourage deviation. The rhythms of the year felt dependable. They had never failed decisively within living memory.

Transportation schedules, energy grids, supply chains—each system rested on the quiet assumption that time itself behaved predictably. Not perfectly, but within limits narrow enough to plan around. Deviations were absorbed locally. Corrections were applied incrementally. Nothing about daily life suggested that the frameworks underpinning it required reconsideration.

Inside, the numbers accumulated.

Elena sometimes imagined the dataset as a ledger—one no one had intended to keep, but which now existed simply because stopping would feel dishonest. The measurements arrived automatically, logged with timestamps and identifiers that stripped them of narrative. Each entry was small. Harmless. A fraction of a millimeter. A fraction of a millisecond. Nothing that couldn't be explained, contextualized, or smoothed away.

But together, they told a story that no single graph could summarize without sounding alarmist.

The ledger grew. The system had been built to continue. Shutting it down would require justification. Maintaining it required none. The experiment had become its own argument for persistence.

That was the irony. The more accurate the measurements became, the harder it was to talk about them.

Precision eliminated spectacle. Each improvement narrowed uncertainty without widening imagination or urgency. The data became cleaner even as its implications grew harder to articulate. Any attempt to frame the results in broader terms risked sounding speculative, even when grounded entirely in observation.

"What do you even say?" Jonah asked one evening as they shut down their terminals. The lab lights dimmed slightly as systems entered standby, a routine response to inactivity. "The Moon's doing exactly what physics says it should. No surprises. No anomalies. No drama."

Elena leaned back in her chair, eyes still on the screen even as the display faded. "You say that the systems we built assumed stability," she replied. "And stability isn't the same thing as permanence."

Jonah smiled faintly. It was the kind of smile that acknowledged a point without embracing it. "That won't fit in an abstract."

"No," she said. "But it fits in reality."

They gathered their things slowly, neither in a hurry to leave. Outside the lab, evening settled into the familiar pattern of departure and return. Traffic thickened, then thinned. Lights came on in offices and apartments, each one an implicit vote of confidence in the continuity of systems that had never demanded justification.

Inside the building, the servers continued their quiet work. The ledger remained open. Another entry would be added before morning, indistinguishable from the last except by its position in time.

The system required no acknowledgment to proceed. Continuity was its only measure of success.

Nothing about the day had required alarm. Nothing about the data demanded instant action. The numbers lay where they always did—orderly, compliant, uninterested in consequence. But familiarity didn't ease the pressure. It sharpened it:
the quiet knowledge that precision could still leave meaning unresolved.

And yet, Elena felt the familiar tension settle in—the awareness that accuracy and relevance were drifting apart, that understanding something precisely didn't guarantee knowing how to live with it.

The world outside didn't need certainty to function. It needed assumptions that worked often enough to be trusted. The question that lingered, unspoken, was how long that trust could be sustained once precision outpaced the stories designed to contain it.

For now, the systems held.

And so the numbers kept arriving.

∝ ∝ ∝

The margin of error continued to shrink.

Not dramatically. Not in a way that invited headlines or demanded explanation. It narrowed the way mature measurements always do—incrementally, almost politely, as instruments improved and procedures refined themselves through repetition. Each year's contribution was modest. Each revision justified. Nothing about the process suggested urgency.

What changed wasn't the data.

What changed was the room around it.

At first, the uncertainty had been whether the measurement itself was correct. Early discussions lingered on instrumentation, alignment, calibration. Questions were practical and contained. Had the signal been clean? Were atmospheric effects fully accounted for? Did independent stations agree within tolerance? These were familiar problems with familiar solutions, and they yielded to careful work.

Once confidence in the measurement solidified, the uncertainty shifted.

Then, the question became whether the rate was constant. Analysts compared years, then decades, looking for curvature where none appeared. Models were stress-tested against the expanding record over time.

Assumptions were challenged, then reinstated. The line behaved itself. If it bent, it did so within expectations. Stability reassured.

As time accumulated, the question shifted again.

It became whether the implications were meaningful. Not incorrect—no one suggested that—but relevant. Was the trend significant beyond its narrow domain? Did it belong in the same category as other long-term processes already absorbed into planning models? Or was it simply another slow variable, measurable but safely ignorable, a curiosity rather than a constraint?

Those conversations were quieter. Less technical. More careful with language.

Now, quietly, without anyone voting on it, the question had become something else entirely.

Not *is this happening?*

That had been settled long ago.

But *how long can we pretend this doesn't matter?*

No single moment marked the transition. There was no memo, no revised charter, no change in formal mandate. The shift occurred the way institutional questions often do—by exhaustion rather than decision, through repetition, deferred arguments, and the slow normalization of uncertainty over time. Each familiar answer, repeated often enough, began to sound less like resolution and more like avoidance.

The data had not forced the change. The accumulation had.

Each additional year made dismissal harder to justify without explicitly naming it as such. Each refinement reduced the space in which uncertainty could hide. The question didn't arrive as an accusation. It arrived as a residue—the leftover unease that remained after all reasonable explanations had been applied.

And that was precisely the problem.

The question no longer belonged to physics.

Physics had done its work. It had described the system accurately, repeatedly, within error margins that left little room for reinterpretation.

What remained wasn't a failure of theory or measurement, but a failure of categorization. The issue no longer fit cleanly inside a discipline accustomed to answering questions by narrowing them.

This question asked something broader.

It asked about relevance. About thresholds. About how much change could be absorbed before assumptions quietly expired. It asked whether a system could remain technically understood while becoming practically misaligned.

And because it didn't belong to physics, it had no obvious home.

That was the most unsettling measurement of all.

∝ ∝ ∝

Signal and Noise

The inbox filled faster than anyone could read it.

That wasn't unusual. It was the nature of modern research—alerts, preprints, revisions, funding notices, system warnings, collaborative pings, calendar invitations stacked on top of one another until urgency flattened into sameness. Everything arrived marked important. Very little was. Priority became a visual style rather than a judgment, a flag applied automatically by systems that couldn't distinguish between consequence and activity.

Elena learned to recognize the sound of certain messages before she opened them. The short, clipped subject lines from automated systems reporting routine completions. The carefully neutral language of committee correspondence, engineered to convey obligation without direction. The occasional late-night email from a colleague who should have been sleeping but wasn't, its tone just informal enough to suggest that it might require a response before morning.

She triaged instinctively. Delete. Archive. Flag. Forward.

The decisions took seconds, guided less by content than by familiarity. Some messages demanded attention because they always had. Others were deferred because they always could be.

The lunar ranging data didn't announce itself.

It arrived quietly, folded into weekly summaries and quarterly reviews, buried among datasets that spiked and jittered and demanded attention. Solar activity surged and dipped. Atmospheric circulation shifted unpredictably. Ice mass redistributed itself in fits and starts, registering abrupt changes that triggered notifications, follow-up requests, urgent discussions. Models adjusted. Confidence intervals widened. Thresholds were revisited. Baselines rechecked repeatedly. Assumptions were flagged, annotated, and carried forward, their implications deliberately deferred in the language of ongoing assessment. Earth rang and flexed and breathed under the weight of its own complexity, its signals noisy, volatile, and impossible to ignore.

Compared to that, the Moon was boring.

It moved the same way every year, in the same direction, at the same rate. No surprises. No drama. No sudden excursions beyond expectation. The line traced by the measurements was clean and unbroken, its slope so gentle it could disappear entirely when plotted alongside more energetic phenomena. There were no spikes to investigate, no thresholds crossed, no alerts triggered by design.

Consistency became camouflage.

The lunar data asked nothing of the people receiving it. It didn't interrupt meetings or demand explanations. It could be skimmed, acknowledged, and set aside without consequence. In a system trained to respond to deviation, stability registered as absence.

"Nothing's changed," someone said during a cross-disciplinary briefing, gesturing vaguely at a screen crowded with colorful plots. "At least, nothing new."

The statement wasn't challenged. In context, it was true.

The Moon was doing exactly what it had been doing before, and before that, and before that. The models held. The predictions remained valid. There was no reason to linger.

Elena watched the lunar line sit in the corner of the display, a pale thread among louder signals. It didn't pulse or oscillate. It didn't flare. It held its position, indifferent to thresholds and alerts. Present without insistence, persistent without urgency. It didn't carry error bars large enough to provoke discussion. It was there because it had always been there, rendered small by scale rather than insignificance.

The meeting moved on.

Slides advanced. Questions clustered around datasets that resisted explanation, around behaviors that defied expectation and therefore demanded interpretation. Attention followed volatility. It always did. The human eye, trained by evolution and reinforced by institutions, gravitated toward change, toward movement that suggested danger or opportunity.

The Moon offered neither.

Its signal blended into the background hum of known quantities, its importance diluted by familiarity. Precision, in this context, worked against relevance. The more predictable the measurement became, the less claim it had on attention.

That was its strength.

The data survived untouched by the filtering mechanisms designed to manage overload. It passed through inboxes and reports without triggering escalation, accumulating quietly, unchallenged and unprotected by the very systems meant to surface what mattered most.

Signal drowned in steady noise.

And in a world calibrated to react to disruption, steadiness became invisible.

∝ ∝ ∝

Noise was what people responded to.

Noise arrived abruptly and demanded interpretation. It came with spikes and thresholds and breaking points—numbers that crossed lines drawn in advance, values that violated expectations loudly enough to trigger alarms. Noise made headlines because it was easy to point to, easy to dramatize, easy to frame as a deviation from normal. It justified emergency meetings and executive summaries precisely because it disrupted existing narratives.

Noise could be ranked. Prioritized. Assigned owners.

Noise fit neatly into systems designed to manage risk, because it announced itself as risk.

The Moon offered none of that.

Its signal didn't spike or collapse. It didn't oscillate or reverse direction. It didn't cross thresholds because no one had defined thresholds for something that moved so slowly. Its behavior resisted the visual language of urgency. On dashboards crowded with color, it appeared as a thin, muted trace—present, accurate, and easy to overlook.

Its steadiness allowed it to blend into the background of institutional life.

Each individual measurement was trivial. A fraction of a millimeter here, a fraction of a millisecond there. Too small to feel consequential, too incremental to provoke response. Each annual change could be dismissed without distortion as negligible. The trend existed only in accumulation, and accumulation required patience the system rarely exercised.

The human mind wasn't built to feel urgency in millimeters.

It was built to notice discontinuities—to react to sudden shifts, sharp edges, abrupt failures. It was built to respond to storms, not tides. Gradual change failed to trigger a response. In environments saturated with data, attention flowed toward what resisted explanation, not what conformed to it too well.

The Moon conformed beautifully.

"What's the signal-to-noise ratio on this?" a climatologist asked once during a review, the question delivered with a tone that balanced curiosity and humor. Half-joking. Half-serious. A way of acknowledging the data without elevating it.

Elena answered without looking up. "Infinite."

That earned a laugh.

The word registered as exaggeration, a clever reply rather than a precise description. It sounded like a punchline because it contradicted instinct. In practice, signal-to-noise ratios were never infinite. Every measurement carried uncertainty. Every instrument introduced distortion. The phrase felt rhetorical, not literal.

It wasn't.

There was no noise in the lunar signal in the sense people meant it. No random fluctuation masking the trend. No volatility to be averaged out. The movement was clean, persistent, and directionally consistent over timescales long enough to dwarf individual careers. What uncertainty remained lay not in the measurement, but in what to do with a signal that refused to misbehave.

The problem wasn't extraction. It was interpretation.

In systems designed to surface what changed suddenly, steadiness registered as absence. The Moon's signal didn't compete for attention because it never raised its voice. It asked no questions of those receiving it. It persisted by remaining unremarkable, logged, archived, and deferred without remark. It made no claims that demanded rebuttal or refinement beyond notation or escalation. It could be acknowledged without being engaged.

That was its strength.

It passed through filters meant to suppress distraction and emerged unchanged on the other side, neither amplified nor flagged. It accumulated quietly, immune to the mechanisms that elevated other data by making it loud. The very qualities that made the measurement scientifically valuable made it institutionally invisible.

Noise provoked response.

Silence endured.

And in an environment calibrated to react to disruption, endurance didn't register as importance—it registered as background.

∝ ∝ ∝

There was a meeting—there were always meetings—where someone suggested bundling the lunar data into a broader report, something that would give it context, make it easier to digest. The suggestion came midway through the agenda, after routine updates and before scheduling conflicts absorbed what little attention remained.

It was framed as a practical concern. Not opposition. Not skepticism. Just a question of presentation.

"We don't want it standing alone," the person said. "People misinterpret isolated trends."

Several heads nodded. The logic was well-known, rehearsed enough to feel self-evident, familiar, procedural, institutional. Context protected against exaggeration, insulating interpretation from reaction and keeping concern proportional to evidence. Framing, reliably and deliberately, prevented misunderstanding as much as error. No one in the room objected to the principle itself. The danger of decontextualized data was something they all acknowledged, at least in theory.

Elena leaned forward slightly. "Or," she said, "they finally see it."

The room paused—not long enough to feel confrontational, but long enough for the comment to register. Someone smiled politely. Someone else looked down at their notes. The moment passed without being resolved, absorbed into the quiet mechanics of discussion.

The suggestion was tabled.

Not rejected. Just deferred.

The chair summarized it carefully. "Let's revisit this again once we have the full comparative analysis." A reasonable condition. A familiar one. It implied progress without requiring commitment.

Deferred decisions had a way of becoming permanent. They accumulated quietly, layered beneath more urgent items. Each postponement arrived with a rationale that felt sufficient at the time. Review became optional, then unlikely. Memory thinned where accountability should have been. What was delayed longest began to feel settled.

They didn't disappear. They simply stopped occupying active space. Once tabled, they migrated into future agendas that never quite arrived, replaced each cycle by more immediate concerns. No one felt responsible for retrieving them. No one felt authorized to force the issue. In subsequent meetings, the lunar data appeared as expected—present, accurate, folded neatly among other results. It no longer stood alone. Nothing had been done to preserve its isolation. Integration happened by default.

Bundling was easier than emphasis. Context diluted without distorting. The data remained intact, but its prominence softened as it shared space with variables that behaved more dramatically, more responsively.

No one could point to a moment when the decision not to highlight it was made.

It simply happened.

And because it happened through deferral rather than refusal, it carried no trace of intent. There was nothing to contest, nothing to reverse. The process had moved forward exactly as designed—by waiting.

∝ ∝ ∝

Late one evening, long after the building had emptied, Elena pulled up two datasets side by side. The corridors outside the lab were dark, lights dimmed to their overnight settings, motion sensors long since satisfied that no one remained. The hour had slipped past productivity into something quieter, where work continued less out of obligation than because stopping felt arbitrary.

On one screen, climate variability: jagged lines stacked over one another, overlapping curves in competing colors, uncertainty bands that ballooned and contracted with each new model run. The data resisted flattening. Each attempt to summarize it revealed new complexity—regional divergence, nonlinear feedbacks, thresholds that appeared stable until they weren't. The plot looked alive, restless, unwilling to settle.

On the other screen, the lunar distance record: a single, unwavering ascent.

No spikes. No oscillations. No competing trajectories. Just a clean line rising gently across decades, its slope so consistent it seemed almost theoretical. The software rendered it in the same neutral palette as everything else, but it might as well have been grayscale compared to the visual noise beside it.

She stared at them longer than she meant to.

The climate data was louder, more chaotic, more demanding. It crowded the eye, pulled attention toward its extremes, insisted on interpretation. It told a story of systems under stress, of feedback loops colliding, of influences layered so densely that separating cause from effect felt like an act of judgment rather than analysis. It was hard to read and harder to explain, and precisely for that reason, it commanded respect.

The lunar data told a simpler story.

It simply refused to perform.

It didn't dramatize itself. It didn't fluctuate in ways that invited speculation. It didn't offer inflection points or sudden reversals that could be framed as turning moments. It advanced at the same pace it always had, indifferent to the surrounding complexity.

She adjusted the scale, just to be sure. The line remained smooth.

The contrast was unsettling. One dataset demanded conversation by virtue of its disorder. The other asked for patience, and patience wasn't something the system rewarded.

"You're still here," said Jonah's voice from the doorway.

She hadn't heard him arrive. He leaned casually against the frame, jacket slung over one shoulder, as though he had already decided he wouldn't stay long.

"Just comparing," she said.

He stepped closer, eyes moving from one screen to the other. He didn't need explanation. The juxtaposition spoke for itself.

"It's funny," he said after a moment. "If that line were jagged, people would take it more seriously."

"If it jumped," Elena said, "they'd call it a crisis."

Jonah nodded. "But because it's smooth…"

"Because it's smooth," she finished, "it gets treated like a constant."

They let that sit between them. The truth of it felt obvious once spoken, and yet oddly fragile, the kind of realization that could evaporate in daylight under the pressure of schedules and agendas.

The lab hummed quietly around them. Servers cycled. Cooling fans whispered. Somewhere deeper in the building, automated systems continued their routines, processing inputs, storing outputs, maintaining continuity without regard for interpretation.

They stood there in silence, listening to the low, steady sound of equipment that would keep running whether anyone watched it or not.

The machines didn't care which screen drew more attention.

They only continued.

∝ ∝ ∝

The trouble with noise was that it trained attention poorly.

It rewarded immediacy. It conditioned people to scan for spikes, to react to deviation, to associate importance with disruption. Over time, attention learned to equate motion with meaning and volatility with relevance. Systems built to monitor change became increasingly sensitive to what fluctuated, and increasingly blind to what persisted.

Stability came to feel like safety—it didn't trigger warnings. Anything that moved slowly enough, or steadily enough, fell out of the picture. It was simply ignored. It was simply passed over, filtered out by habits of perception that had been reinforced for years.

The Moon exploited that bias perfectly.

It didn't interrupt workflows or challenge expectations. It didn't produce alerts or cross thresholds that had been drawn for other purposes. It never arrived labeled urgent. Its signal didn't force its way into conversations already crowded with louder claims.

It didn't escalate.

Each measurement differed from the last by an amount too small to provoke concern on its own. Taken individually, the changes were dismissible—variations that could be attributed to calibration, rounding, or context. Taken together, they formed a trend that required patience to recognize and discipline to acknowledge.

It didn't demand interpretation.

There was no ambiguity to resolve, no puzzle to solve. The physics was settled. The measurement repeated itself with quiet insistence, offering no resistance to explanation. It asked only to be noticed over time, and time was precisely what attention had been trained not to grant.

It simply continued.

Somewhere in the layers of analysis and reporting and review, the signal remained intact. It was still there, clean and undeniable, preserved faithfully by systems designed to record rather than judge. The data survived its own invisibility. It waited, not for discovery, but for a shift in perspective—for someone to decide that constancy wasn't the same thing as irrelevance.

That decision didn't belong to instruments or equations. It belonged to the people who interpreted what stability meant.

Elena closed the files and shut down her workstation. The screens went dark in the familiar sequence, each window disappearing. Nothing about the act felt consequential. It was the end of a workday like any other.

Tomorrow the data would update again. Another pulse of light. Another return. Another fraction of a millimeter added to a line that refused to be dramatic. The system would record it automatically, append it to the ledger, and move on.

The noise would continue to shout.

It would flare and spike and demand explanation. It would arrive with urgency, insist on response, justify attention by refusing to be ignored. It would dominate meetings and dashboards and headlines because it fit the shape of what people had learned to notice.

The signal would continue to speak.

Quietly. Precisely. Without interruption or emphasis. It would say the same thing it had always said, in the same way, at the same pace. It would make no effort to compete.

And the difference between them would only grow harder to explain.

The habits built around noise grew stronger—training attention to value what changed quickly and discount what endured.

∝ ∝ ∝

The Longest Experiment

No one ever officially named it that.

There was no ceremony, no proposal circulated for approval, no vote recorded in minutes. The phrase appeared first in passing, buried in a footnote of a review paper that summarized three decades of lunar laser ranging results. It wasn't meant to carry weight. It functioned as shorthand, a convenient way to gesture at duration without interrogating it. An experiment that had gone on for a long time needed a phrase that acknowledged that fact without asking what it implied.

The authors didn't linger on the wording. Neither did the reviewers.

The name spread slowly, the way informal terminology often does. It appeared again in a conference slide, then in a seminar talk, then in a grant justification where space was limited and history had to be compressed into a sentence or two. Each time, it arrived without emphasis, treated as a descriptive convenience rather than a conceptual claim.

By then, the work had outlived fashions in science. Paradigms had shifted. Techniques once considered cutting-edge had become standard, then obsolete. Entire subfields had emerged, matured, and fragmented again. Funding priorities rotated, driven by discoveries elsewhere, by political winds, by crises that demanded faster returns.

The experiment remained.

It had also outlived funding cycles themselves. The original justifications that secured support were no longer cited. New ones had been written, each reframing continuity as operational necessity rather than mere inertia. The budget lines persisted because removing them would have required explanation.

Governments rose and fell. Administrations changed. Agencies were reorganized, rebranded, defunded, revived. Names changed. Logos changed. Oversight structures shifted. The experiment passed through these transitions without comment, its presence carried forward by bureaucratic momentum rather than deliberate choice.

The mirrors remained.

Their status was implicit, carried forward without review. Absence of objection functioned as approval.

Persistence required no affirmative act.

They didn't require permission to keep existing. No renewal application governed their presence on the lunar surface. No review committee evaluated their ongoing relevance. They occupied their positions without protest or justification, unresponsive to the political frameworks that had made their deployment possible.

Every night, somewhere on Earth, a laser fired.

Sometimes from a mountain observatory, dome rotating slowly under clear air. Sometimes from a facility few people outside the field could name, its coordinates familiar to specialists and systems that automated the work.

The exact location didn't matter. The geometry did. Earth rotated. The Moon moved. The timing windows opened and closed.

Light left Earth, crossed the distance, struck the array, and returned. The signal was weak, but sufficient. It always had been. Time closed the loop, translating distance into delay, motion into accumulation.

No single person witnessed the entire process anymore. The firing, the detection, the recording, the archiving—all occurred through systems designed to function without supervision. Human involvement had shifted from participation to oversight, from attention to trust.

The experiment no longer asked a question; that phase had passed quietly, without announcement. The original motivations—testing relativity, refining orbital mechanics, improving models—had been answered well enough to satisfy their framers. What remained wasn't inquiry, but continuation.

It simply collected answers.

Night after night. Year after year. The system had been built to continue, and no one had chosen to stop it. The experiment persisted not as a project with a goal, but as a fact of the scientific landscape—present, reliable, and increasingly difficult to situate within the language of purpose.

There was a peculiar comfort in that, at least at first.

In a world increasingly defined by uncertainty—by systems too complex to model cleanly, by data that argued with itself—the lunar returns felt reassuring in a way few other measurements did. They behaved. They arrived on schedule. They produced results that aligned neatly with expectations established long before computational power made testing those expectations routine.

The Moon didn't surprise anyone.

That alone made it valuable. While other datasets demanded interpretation, negotiation, and caveats, the lunar returns offered confirmation. They validated equations written decades earlier, in an era when confidence had been earned through scarcity rather than abundance.

Each successful return reinforced the sense that at least some parts of the universe remained legible, orderly, and compliant.

In meetings, the data carried no tension. It didn't provoke debate or invite challenge. It could be referenced briefly and moved past without fear of omission. In a research environment saturated with provisional results and contested models, the lunar measurements felt like solid ground.

"This one just works," someone said during a budget review, when the line item came up again.

The phrasing was casual, almost affectionate. It implied reliability without requiring defense. "Low cost. High confidence."

No one objected.

There was nothing to object to. The experiment didn't compete with newer initiatives that promised faster returns or greater visibility. It didn't demand resources that might otherwise be allocated elsewhere. Its infrastructure already existed. Its procedures were automated. Its outputs were predictable.

It didn't produce surprises.

That absence wasn't perceived as an imperfection. In a climate where unexpected results often carried political or administrative risk, predictability was deemed like virtue. The lunar returns posed no awkward questions. They didn't complicate narratives or destabilize planning documents. They confirmed what was already believed, and did so quietly.

The experiment became a fixture—something counted on precisely because it never asked to be counted.

Its presence in budgets and schedules felt less like a decision than an inheritance. Each year it was carried forward by default, justified not by ambition but by continuity. Removing it would have required explanation. Maintaining it required none.

And so it continued.

Night after night, a pulse of light crossed the distance and returned. Another data point joined the ledger. Another confirmation was logged.

The system functioned exactly as designed, offering reassurance through repetition.

∝ ∝ ∝

Elena encountered the phrase again years later, this time in a lecture given to first-year graduate students. The room was crowded in the way introductory courses always were, chairs filled with people still learning how to listen for what mattered. Laptops were open, notebooks half-used, attention divided between the speaker and the habit of recording what might be important later.

Most of the students assumed relevance could be identified by emphasis. Few knew how often significance arrived without announcement.

The speaker used the phrase casually, almost affectionately, as if referring to an old instrument still ticking away in the corner of the lab—reliable, familiar, and not especially demanding of attention.

"The longest experiment in human history," he said, gesturing to a slide of the Moon, bright and untroubled against a black background. "Started in 1969. Still running."

The line drew a few smiles. It sounded impressive without sounding heavy. A curiosity, a point of pride. The kind of fact that fit neatly into a lecture without interrupting its flow.

He moved on briefly, then circled back, clicking through slides that compressed decades into bullet points. Names flashed by. Dates. Diagrams simplified until their origins were barely visible. The mirrors were mentioned only in passing, their role assumed rather than explained.

A student raised a hand.

"What's the hypothesis?"

The question was earnest, framed the way students were taught to frame questions—clean, contained, answerable. It presumed an experiment with a beginning, a middle, and an end.

The speaker smiled. "Originally? That we could measure the distance precisely."

A reasonable answer. Safe. Historical. It anchored the work firmly in the past, where hypotheses could be resolved without consequence.

"And now?" the student asked.

The smile lingered a moment too long.

"Now we're measuring change."

The words were chosen carefully. Neutral. Accurate. Unthreatening. They didn't suggest urgency or implication. They sounded like a natural evolution of the original goal, an extension rather than a transformation.

The student nodded, satisfied, and wrote something down.

Elena watched the pen move across the page, wondered what the student had recorded. A phrase. A date. A margin note. A definition stripped of context. Probably nothing that would trouble them later.

The lecture continued. Slides advanced. The phrase receded into the background, absorbed into the structure of the course. It would be repeated, perhaps, on an exam. It might appear again in a seminar, introduced with the same tone of mild admiration.

By the time the lecture ended, the phrase had done its work. It had conveyed continuity without weight, longevity without obligation. It framed the experiment as an artifact of history rather than an active presence, something inherited rather than interrogated.

Elena gathered her things and left with the students, blending into the flow of people headed for their next commitments.

The Moon remained on the screen behind them, unchanged, its image already giving way to the next slide.

The longest experiment in human history.

It sounded complete.

∝ ∝ ∝

The problem with experiments that never end is that they stop feeling like experiments at all.

They become infrastructure.

They inherit the presumption of necessity.

Their operation shifts from question to condition.

They fade into the background, assumed, trusted, unexamined. No one asks whether they should continue, because stopping them would feel arbitrary, almost dishonest. You don't turn off a clock just because you don't like what time it's becoming.

And clocks were, in a way, what the mirrors had become.

They marked distance with time. They logged motion without drama. They counted without commentary.

They didn't care who was watching.

The mirrors had outlived their original questions. The proposals that justified their installation had yellowed in archives, their language careful, provisional, optimistic in a way that now felt almost naïve. The scientists who had argued for them had retired, then departed this life. Their names survived only in citations, abbreviated and stripped of context, attached to results that no longer required explanation.

The mirrors remained.

They were maintained the way one maintains bridges or power lines—failure would be unacceptable. A broken mirror would raise questions. A missing data point would invite scrutiny. Continuity, by contrast, required no defense.

So the measurements continued.

Technicians recalibrated instruments. Software was updated. Clocks were synchronized to ever-finer standards. Each improvement folded seamlessly into the existing record, smoothing the past rather than disrupting it. The dataset grew denser, more confident, more resistant to dismissal.

At no point did anyone decide that the experiment was finished. It simply became easier to keep it running than to explain why it should stop.

Funding agencies classified it as legacy support. Administrators listed it among ongoing obligations, the way one lists utilities or maintenance contracts. Graduate students learned how to handle the data as a matter of technique, not curiosity. It was there. It was reliable. It was used.

No one stood at the center of it anymore.

That absence mattered. Experiments with clear endings demand interpretation. They force conclusions. They require someone to say *this is what it means*. An experiment without an ending can afford patience. It can wait for relevance to arrive on its own.

Over time, the distinction between observation and assumption blurred. The Moon's distance became a known quantity. It was measured so consistently that any change would have felt theoretical. Motion stretched thin across decades does not look like motion at all. It looks like stability with a footnote.

Until it doesn't.

The mirrors didn't care when that threshold was crossed. They didn't register surprise. They didn't adjust their behavior in response to debate or dismissal. They reflected light, measured time, and waited for the rest of the system to decide what those measurements were worth.

They had no concept of significance.

That burden belonged entirely to the people who had stopped thinking of them as experimental at all.

∝ ∝ ∝

In private moments, Elena sometimes imagined future researchers scrolling back through the dataset, long after she was gone, long after the people in her meetings had been replaced by others with the same titles and different faces and incentives. She imagined someone zooming out far enough to see not sixty years, but a century. Two.

At that scale, individual measurements would vanish into continuity. Daily effort would collapse into trend.
What once felt deliberate would read as inevitable.

She imagined the interface would look different by then. Cleaner. More abstract. The early years compressed into a narrow band, the later ones dominating the screen. The annotations would have multiplied—layers of commentary added by people who had never met the original architects, citing papers written by authors who had once been graduate students themselves.

The names would blur. The institutions would persist.

The line would still rise.

Slowly. Relentlessly.

She imagined the researcher pausing, fingers hovering over a control, deciding whether to keep zooming. There would be a moment, she thought, when the scale stopped feeling safe. When the slope, so modest year to year, became undeniable when viewed whole. Not dramatic. Not catastrophic. Just continuous.

It wouldn't curve sharply. It wouldn't spike. It wouldn't announce itself as a crisis. It wouldn't justify urgency on its own.

It would simply persist.

Its presence would feel informational rather than directive.

Elena had learned, over time, that persistence was more dangerous than shock. Sudden changes forced reaction. They demanded explanation. They mobilized attention. Gradual ones invited accommodation. They encouraged people to adjust expectations instead of behavior.

A line that rose slowly enough could survive generations of scrutiny without ever provoking a decision.

She imagined the future researcher checking metadata, confirming calibration notes, ruling out drift. The same process she had followed. The same reassurance. The same quiet unease when every alternative explanation failed to fully account for the trend.

By then, the context would be different. Other problems would have claimed the foreground. Other emergencies would have justified delay. The dataset would be treated as historical evidence rather than a warning, its early entries already old when Elena was young.

Still, the line would be there.

It wouldn't care who first noticed it. It wouldn't reward insight or punish neglect. It would remain indifferent to the moment when someone finally decided that persistence itself constituted a message.

And because of that persistence, it would eventually outlast denial.

Not by force. Not by argument. But by remaining visible long enough that disbelief exhausted itself. Long enough that the cost of ignoring it exceeded the cost of explaining it away.

Elena sometimes wondered whether that was how truth usually worked—not as revelation, but as attrition.

∝ ∝ ∝

The last meeting of the quarter ended the way most of them did: with action items assigned, language softened, and the unspoken agreement that nothing fundamentally needed to change yet.

The room emptied gradually. Chairs slid back into place. Screens went dark. Someone joked about coffee as a substitute for progress. The official minutes would summarize the discussion in neutral terms—milestones reviewed, concerns noted, next steps deferred. Nothing in them would suggest that anyone had felt the weight of what they were discussing, or that the numbers on the screen had lingered in anyone's mind after the projector shut off.

As people gathered their things, Jonah leaned toward Elena. "You ever think about how long this thing's been running?"

"All the time," she said.

"And how long it'll keep running?"

She nodded. "Longer than us."

Jonah considered that. "That's not usually how experiments go."

"No," Elena said. "But this one isn't really ours anymore."

He didn't argue. There was nothing to add. Experiments implied ownership—hypotheses, endpoints, conclusions. This had slipped past all of that. It had become something closer to an obligation, a background condition of the world they worked in.

They left the room together, the lights switching off automatically behind them. The building's systems didn't pause to register their departure. Climate control adjusted. Security sensors armed themselves. Somewhere, a server logged the meeting's end time and moved on.

Outside, the night had already settled into its familiar arrangement. The city moved at its own scale—traffic lights cycling, windows brightening and dimming, people measuring their evenings in hours rather than decades.

Somewhere far above, the Moon hung where it always had, carrying with it no sign of the measurement being taken, no hint of the ledger growing longer with each passing night.

No mark.

No signal.

No acknowledgment.

The photons left Earth, struck the mirrors, and returned. The clocks advanced. The distance increased by an amount too small to feel, too small to matter in any immediate sense.

The experiment continued.

It didn't care whether meetings ran long or short. It didn't register softened language or postponed decisions. It didn't know what a quarter was, or why its end carried significance.

It didn't ask whether humanity was paying attention.

It only recorded whether humanity was still there to read it.

∝ ∝ ∝

Something has been observed continuously, precisely, and without interruption for more than half a century.

That alone was unusual. Most measurements frayed at the edges. Instruments failed. Funding lapsed. Attention wandered. Datasets accumulated gaps that later had to be bridged with inference and assumption. This one didn't. It endured upgrades, revisions, changing standards, and the quiet churn of institutions without ever losing continuity.

What had begun as an experiment became a record.

It survived because it asked so little. A pulse sent. A pulse returned. Time measured. Distance calculated. Night after night, year after year, the same exchange repeated with mechanical patience. No intervention required. No interpretation demanded.

What comes next is not about measurement.

This was the point where observation usually gives way to urgency. Instead, it produced language.

The numbers don't need refinement. The error bars are narrow enough. The clocks are accurate enough. The mirrors are still answering. There is no technical threshold left to cross that would make the result clearer than it already is.

What comes next is about what people decide to do with certainty when certainty becomes inconvenient.

Certainty is often treated as resolution, but it rarely behaves that way. When it confirms expectation, it disappears into the background, reinforcing decisions already made. When it contradicts them, it becomes a problem to be managed rather than a fact to be confronted.

In practice, certainty does not compel action. It invites negotiation.

Negotiation reframes facts as positions.

Positions become agendas.

Agendas acquire process.

Process stretches time.

Time dulls urgency without ever disproving the original claim.

People argue about context. They argue about relevance. They argue about whether the certainty applies *yet*, or whether it can be deferred in favor of competing priorities. They ask whether a trend measured across decades must matter now, or whether it can remain provisional simply because its consequences unfold slowly.

Institutions are particularly skilled at this. From here on, the experiment was no longer something observed. It was something managed.

They absorb certainty and distribute its implications across committees, reports, and timelines until no single person feels responsible for responding to it. The data remains intact. The response dissolves.

There is comfort in that diffusion. Certainty that demands no immediate change is easier to live with than uncertainty that imposes a choice, especially when responsibility can be shared. A result that cannot be ignored but also cannot be acted on cleanly can linger indefinitely, acknowledged in principle and sidelined in practice.

The mirrors don't participate in that negotiation.

They don't distinguish between a result that is welcome and one that is awkward. They don't know which conclusions are affordable and which are not. They provide the same answer regardless.

What changes is everything around them.

Certainty becomes something to be contextualized, reframed, or postponed. It is accepted as real, but treated as premature. Not wrong—just early. Not alarming—just incomplete. Not actionable—yet.

Yet can stretch for a very long time.

Long enough for responsibility to shift. Long enough for those who first recognized the certainty to be replaced by others who inherited it without the memory of discovery. Long enough for the measurement to feel less like a finding and more like a condition of the world.

At that point, the question is no longer whether the certainty exists.

It is whether anyone is willing to treat it as binding.

PHASE 2—THE DATA STOPS BEING NEUTRAL

The conference room was too bright for the hour.

The overhead lights had been set for midmorning productivity, not for the hour just past dawn when bodies were still negotiating the transition from sleep to obligation. The glare flattened shadows and erased the sense of time outside. It made the room feel sealed, self-contained, as though whatever happened inside it wouldn't easily disperse.

Dr. Lena Morales stood at the screen with her arms folded, not presenting—waiting. She had learned the difference over years of similar meetings. Presenting meant guiding attention. Waiting meant letting the data speak first, forcing everyone else to confront it without narrative assistance.

Around the table, coffee went untouched. Steam rose briefly, then faded. No one had opened a laptop yet. That alone felt wrong. Laptops were armor, a way to insert distance between oneself and the moment. Their absence suggested that whatever was about to be discussed couldn't be safely multitasked.

"Run it again," said Marcus Reed, finally.

His voice carried the calm of someone accustomed to issuing instructions that machines obeyed. Not authority exactly—expectation. Systems, when properly designed, did what they were told.

Lena didn't argue. She tapped the control.

The graph re-rendered.

The same slope. The same deviation. No spike. No discontinuity. A line that refused to sit where it was supposed to. Nothing about it justified or required escalation at this time. It was persistent enough to register, restrained enough to evade alarm. It wasn't dramatic enough to provoke alarm, but it was precise enough to resist dismissal. The deviation sat inside the frame like a deliberate mark, neither error nor outlier.

"It's within tolerance," said Jun Park, squinting. "Barely, but still inside."

He leaned forward slightly, as though proximity might resolve ambiguity. Tolerance had always been a safe word. It implied containment. Control.

Lena nodded. "That's true."

The room waited for her to add something else. She didn't.

The pause lengthened without anyone claiming it.

Silence began to feel like a position.

No one spoke.

Marcus leaned back in his chair, the metal legs scraping faintly against the floor. "That wasn't a defense."

"It wasn't meant to be."

Dr. Ruth Calder broke the silence from the far end of the table. Her voice was quiet, but it carried. "How many cycles now?"

"Seven," Lena said. "Eight if you count the partial."

Ruth absorbed that without visible reaction. "And all of them agree?"

"Yes."

Jun exhaled through his nose, the sound sharp but controlled. "That's not how noise behaves."

The word landed differently than it had in other rooms, other contexts. Noise had always been the safe explanation—the thing you blamed when patterns appeared prematurely. Noise bought time. But this pattern had survived time.

Marcus rubbed his thumb against the rim of his mug, a small, repetitive motion that suggested he was buying himself a moment. "You're saying the instruments are wrong."

Lena shook her head. "I'm saying they're consistent."

"That's worse," Ruth said quietly.

No one contradicted her.

The room settled into a different temperature—not colder, just heavier. The air felt more deliberate, as though every breath now had to justify itself.

The data on the screen had not changed, but its status had. It was no longer just information. It was pressure.

Marcus glanced at the door, then back at the screen. The gesture was unconscious, a reflex toward exits when things stopped behaving predictably. "We don't escalate on a line."

"No," Lena agreed. "We escalate on what the line means."

Jun finally opened his laptop. The sound of the hinge was sharp in the silence, too loud for what should have been a routine motion. The screen lit his face faintly as he began typing, not notes yet—anchors. Something to hold onto.

"So what are we calling this?" he asked. "Drift? Artifact?"

Names mattered. Everyone in the room knew that. A name transformed observation into category. Category invited precedent. Precedent invited procedure.

Lena hesitated.

"I'm not naming it yet."

Marcus looked up. "Why not?"

"Because once it has a name," she said, carefully, "it stops being neutral."

The words settled slowly, finding purchase. No one rushed to fill the silence that followed. They all understood what she meant, even if none of them wanted to say it aloud.

No one disagreed.

The graph remained on the screen, unchanged, patient. The line didn't react to scrutiny or hesitation. It simply existed, waiting for the room to decide what kind of object it had become.

∝ ∝ ∝

WHAT THE DATA ACTUALLY SAYS (QUIETLY)

The meeting ended without a conclusion, which everyone understood meant it would never really end.

Chairs slid back. Coffee cups were finally lifted, then abandoned again. Laptops closed with the soft finality of people pretending they were done.

Lena was packing up cables when Marcus stopped beside her.

"Don't circulate this yet," he said, low enough that it didn't sound like an order.

"I wasn't going to," she replied.

"That wasn't what I meant."

Lena straightened. "Marcus, if you're asking me to hold raw measurements—"

"I'm asking for time," he cut in. Then, more carefully: "I'm asking for *sequence*."

Across the room, Jun lingered near the window, phone in hand but untouched. Ruth Calder was already gone.

"Sequence for what?" Lena asked.

Marcus hesitated. Just long enough.

"For context."

Lena gave a short, humorless smile. "Context doesn't change slopes."

"No," he said. "But it changes consequences."

Jun turned from the window. "You're talking like this is already a problem."

Marcus looked at him. "I'm talking like it *might* become one."

Jun frowned. "That's not how science works."

Marcus met his eyes. "That's exactly how institutions work."

The word sat between them, heavier than the data ever had.

Jun slipped his phone into his pocket. "Elena Ortiz should see this."

Marcus didn't answer.

Lena watched him carefully. "She already has."

That got his attention.

"She asked for the raw series last night," Lena continued. "No commentary. No framing. Just the numbers."

Jun's eyebrows rose. "And?"

"And she didn't respond."

Marcus exhaled slowly. "That's not like her."

"No," Lena agreed. "It's not."

Silence crept back in—not the tense kind from before, but something more deliberate. A pause with intention behind it.

Jun broke it. "So what happens now?"

Marcus looked around the emptying room, at the screen still glowing with the unhelpful line.

"Now," he said, "we wait and see who blinks first."

Lena frowned. "That's not a scientific answer."

Marcus gave a tired smile. "Phase changes rarely are."

They left the room one by one.

The screen stayed on.

The line didn't move.

∝ ∝ ∝

Lena didn't go home.

She stayed in her office after the building thinned out, lights dimming floor by floor, the custodial carts whispering past like tides. The numbers were already open on her screen—not the new ones, not the slope that refused to behave, but the older files. The ones everyone assumed they understood.

She pulled the Apollo datasets first.

Not out of nostalgia. Out of habit.

Laser ranging. Retroreflectors. Earth-to-Moon distance measured by timing photons that had no reason to lie. Millimeters of precision, accumulated over decades. The kind of data scientists trusted because it didn't *ask* anything of them.

The figures were immaculate.

Distance increasing.

Rate steady.

About 3.8 centimeters per year.

She'd taught this. Everyone had.

Tidal mechanics. Angular momentum transfer. Earth slows, Moon drifts. Clean equations. Closed loops. No mystery.

Nothing in the numbers looked dangerous.

Nothing jumped.

Lena leaned back in her chair and let the familiarity settle.

This wasn't fringe data.

It wasn't disputed.

It wasn't political.

It lived in textbooks. In appendices. In footnotes that never reached the conclusions.

That was the problem.

The data didn't shout.

It didn't even whisper urgently.

It simply continued.

She opened a paper from the late twentieth century, skimmed the abstract, then the conclusion. The language was calm, almost bored. *Expected behavior. Consistent with models. No anomalous deviation detected.*

Filed. Archived. Cited. Forgotten.

Lena closed the document and stared at the wall.

Nothing about the data, by itself, demanded attention.

That was why it survived unchanged for so long.

And that—she was beginning to understand—was what made it dangerous.

∝ ∝ ∝

The argument didn't start as an argument.

It started the way most of them did—someone trying to be helpful.

"You're over-indexing on it," said Paul Henning, tapping the table with a pen he hadn't stopped clicking since he sat down. "That dataset is the most boring thing we have."

Lena didn't look up from the screen. "Define boring."

Paul shrugged. "Stable. Clean. Measured to death. We've been firing lasers at the Moon since Apollo. Millimeter precision. Decades of data. It's a solved problem."

Across the table, Jun frowned. "Solved doesn't mean irrelevant."

Paul turned to him. "It means *settled.* The recession rate's been stable since the late twentieth century. Three point eight centimeters a year. Give or take rounding. That's not controversial."

"No one said it was," Lena said.

"Then why are we talking about it like it's radioactive?"

Ruth Calder finally spoke. "Because it's the foundation."

Paul rolled his eyes. "It's tidal mechanics. Earth loses angular momentum, Moon drifts outward. We teach this to undergrads."

"Yes," Ruth said evenly. "We teach it because it's clean. Because it doesn't scare anyone."

Jun leaned forward. "That's not a scientific criterion."

Paul laughed. "It is if you want funding."

Lena turned the screen toward them. Not the new plot. The old one. A straight, well-behaved line.

"Look at it," she said. "Nothing jumps. Nothing accelerates. Nothing violates expectation."

Paul nodded. "Exactly."

Ruth folded her hands. "Because by itself, it says nothing alarming."

Paul pointed at her. "Thank you."

Jun shook his head. "No. That's the trap."

Paul sighed. "Here we go."

Jun didn't back down. "You're treating 'not alarming' as the same thing as 'not meaningful.'"

Paul replied, "I'm treating it as 'not actionable.' Which is the same thing in practice."

Lena let the silence stretch before cutting in.

"That data," she said, "is the most trusted measurement we have. Laser timing. No models layered on top. Just distance and time."

"So?" Paul asked.

"So if something changes," she said, "it won't announce itself by screaming. It'll announce itself by still looking reasonable."

Ruth nodded slowly. "Quiet data survives scrutiny."

Paul stared at the line, jaw tight. "You're saying we missed something."

"I'm saying," Lena replied, "that nothing about that dataset ever demanded attention."

Jun finished the thought for her. "Which is why it got ignored."

No one spoke after that.

Paul uncapped his pen, then capped it again. "You realize how insane this sounds, right? You're arguing that the problem is that the data behaved exactly as expected."

Lena met his eyes. "Yes."

"That's not how alarms work."

She gave a small, tired smile. "That's how long experiments work."

∝ ∝ ∝

The First Real Question Scientists Ask

Paul pushed his chair back an inch, then stopped himself, as if motion alone might escalate things.

"Okay," he said. "Let's pretend I buy this. Let's pretend boring data is somehow dangerous. What exactly do you think we missed?"

Lena didn't answer right away. She looked at Ruth instead.

Ruth took the cue. "We didn't miss *movement*," she said. "We missed dependency."

Paul blinked. "Dependency on what?"

"On the Moon," Jun said.

Paul let out a short laugh. "The Moon isn't going anywhere."

"No one said it was," Lena replied.

"That's literally the first thing people always jump to," Paul said. "Apocalypse-by-orbit. We're not doing that."

"We're not," Jun said. "That's not the question."

Paul folded his arms. "Then ask it."

Lena turned the screen again, this time not to a graph but to a block of equations—long-term climate models, orbital parameters highlighted in faint yellow.

"The real question," she said, "is what we quietly assume never changes."

Paul leaned forward despite himself.

"These models," Ruth continued, tapping the table with two fingers, "don't re-derive first principles every year. They inherit them."

Jun nodded. "Axial tilt. Seasonality. Long-term climate envelopes. All treated as constants."

Paul frowned. "Because they *are* constant. On human timescales."

Ruth looked at him. "On whose models?"

Lena stepped in before he could answer.

"The Moon doesn't have to leave," she said. "It just has to stop doing what we depend on it to do."

Silence again—but this time it felt different. Less dismissive. More alert.

Paul glanced at the equations. "You're saying the Moon stabilizes Earth's tilt. Everyone knows that."

"Yes," Jun said. "And we know it because we assume it."

Paul's jaw tightened. "That's not a test."

"No," Ruth agreed. "It's a bake-in."

Lena leaned back against the table. "We don't stress-test seasonality against lunar drift. We treat them as solved boundary conditions."

Paul looked up sharply. "Because if they weren't stable, we'd see it."

"Would we?" Jun asked.

Paul hesitated.

Lena pressed. "What would it look like at first? A half-degree wobble over centuries? A shift that never shows up in a funding cycle?"

Ruth's voice dropped. "A change slow enough to be categorized as background."

Paul ran a hand through his hair. "So your concern isn't that the Moon is leaving."

"No," Lena said.

"It's that we built everything," Jun finished, "on the assumption that it stays dynamically relevant forever."

The room felt smaller now. Not trapped. Just… occupied.

Paul stared at the screen. "If that assumption's wrong…"

Ruth didn't let him finish.

"Then the consequences don't announce themselves as disasters," she said. "They announce themselves as *miscalculations*."

No one argued with that.

Paul finally sat back. "So this is the first real question."

Lena nodded. "Yes."

"And it's not about where the Moon is going," he said slowly.

"It's about what breaks," she replied, "if it ever stops holding things still."

∝ ∝ ∝

Paul rubbed his forehead. "Okay. Slow down. You're saying our problem isn't the Moon moving. It's that we stopped asking what depends on it."

"Yes," Lena said. "Because the dependency is boring."

Jun frowned. "That's not fair."

"It is," Ruth replied. "Stability doesn't get revalidated. It gets inherited."

Paul looked between them. "Inherited by what? Be specific."

Lena nodded. "Climate models. Orbital simulations. Long-term habitability projections. Anything that spans more than a few thousand years."

Paul scoffed. "Those are theoretical. No one's making policy off ten-thousand-year projections."

Ruth met his eyes. "We're making *infrastructure* off them."

That landed harder.

Jun leaned forward. "Seasonality assumptions. Agricultural planning. Ice-age boundary conditions. Sea-level equilibrium models."

Paul shook his head. "Those don't hinge on the Moon. They hinge on solar input."

"And axial tilt," Lena said.

Paul opened his mouth, then stopped.

Jun pressed. "You remove or weaken lunar stabilization, tilt becomes a variable. Not chaotic. Just… less well-behaved."

Paul exhaled slowly. "Define 'less well-behaved.'"

"Longer winters in one hemisphere," Ruth said. "Shorter in another. Shifts that take centuries to show up and millennia to settle."

Lena added, "Climate envelopes that drift outside the range we calibrated for."

Paul stared at the table. "But those models already have uncertainty."

"Yes," Jun said. "Bounded uncertainty."

Ruth nodded. "The Moon is one of the things that bounds it."

Silence again. Not the dismissive kind. The calculating kind.

Paul finally said, "So we're not talking about collapse."

"No," Lena replied.

"We're talking about *misalignment*," Jun said. "Predictions slowly decoupling from reality."

Ruth folded her arms. "By the time anyone notices, it looks like model failure. Or bad data. Or regional anomaly."

Paul looked up sharply. "Not like a physical cause."

"Exactly," Lena said. "The Moon doesn't get blamed. The people do."

Paul sat back. "And we don't test this assumption because…"

"Because it's never been wrong before," Jun said.

"And because," Ruth added, "you don't get funded to question constants."

Paul let out a short laugh. "You get funded to refine them."

Lena pointed at the screen again. "That line? Three point eight centimeters a year? It's the reason no one asked the second question."

Paul frowned. "Which is?"

She answered carefully. "What if the most trusted data we have is only telling us *where the Moon is*—not what we quietly asked it to keep doing."

No one spoke.

Jun finally said, "We never stress-tested the system."

Paul nodded slowly. "Because it never screamed."

Ruth finished it. "And boring data doesn't scream."

∝ ∝ ∝

The Critical Analytical Shift

Jun was still staring at the screen when he said it.

"That assumption," he said slowly, "only works if the relationship is linear."

Paul looked up. "Which one?"

"All of them," Jun replied. "Tilt. Torque. Stabilization. We talk about drift like it scales cleanly."

Ruth narrowed her eyes. "It doesn't."

Jun shook his head. "No. It really doesn't."

Paul frowned. "You're saying a small change causes a small effect."

"I'm saying," Jun corrected, "that's what we *assumed*."

Lena turned toward him. "You're thinking about torque curves."

Jun nodded. "And feedback."

Paul sighed. "Okay. Walk me through this before you lose me."

Jun hesitated, then stood, moving to the whiteboard without asking. He didn't draw equations. Just shapes. Arcs. A curve that bent back on itself.

"Axial stability depends on lunar torque," he said. "But not proportionally. It's not a straight line."

Ruth leaned forward. "It never was."

Jun pointed at the curve. "Here's the problem. Small changes in torque don't produce small changes in behavior. They change *predictability*."

Paul crossed his arms. "That's hand-waving."

"No," Lena said quietly. "That's dynamics."

Jun turned back to them. "You don't need the Moon to leave for this to matter. You just need the system to stop behaving nicely."

Paul stared at the board. "Define 'nicely.'"

"Repeatably," Ruth said. "Smoothly. In ways we can average."

Jun nodded. "Before thresholds are crossed, you get chaos. Not explosions. Variance."

Paul's jaw tightened. "Chaos as in…?"

"As in," Lena said, choosing her words, "the same initial conditions stop producing the same outcomes."

Silence.

Paul looked between them. "You're telling me Earth doesn't lose stability when the Moon is gone."

"No," Jun said.

"You lose it," Ruth added, "when you can no longer predict how stability behaves."

Paul swallowed. "That's worse."

Lena nodded. "Because nothing breaks. Nothing announces itself."

Jun capped the marker. "The models still run. The outputs still look reasonable. They just stop agreeing with each other."

Paul exhaled. "So what changes first?"

Ruth answered without hesitation. "Confidence."

Lena finished it. "Forecasts stop lining up. Long-term projections widen. Error bars inflate quietly."

Paul stared at the curve again. "And no one calls that a crisis."

Jun shook his head. "They call it uncertainty."

"And uncertainty," Ruth said, "gets tolerated."

Paul leaned back, hands flat on the table. "So the terrifying part isn't instability."

"No," Lena said.

"It's that predictability decays first."

She looked at the others. No one argued.

"That's not dramatic," Paul said finally.

Ruth met his eyes. "No."

Jun added softly, "That's why it's terrifying."

The whiteboard stood between them, unchanged.

The crack wasn't loud.

But it was there.

∝ ∝ ∝

"What if?" — The Questions That Don't Make Headlines

They didn't write any of it down.

The meeting had technically ended, but no one stood. Chairs stayed where they were. The whiteboard remained untouched, like a witness no one wanted to acknowledge.

Ruth broke the silence first. "What if the tilt varies by a degree?"

Paul looked at her. "It doesn't."

"I know," she said. "I'm not asking *does*. I'm asking *what if*."

Jun rubbed his chin. "One degree wouldn't flip the planet."

"No," Lena said. "But two might stretch seasons."

Paul frowned. "Define stretch."

"Weeks," Jun replied. "Not disasters. Just drift."

Paul let out a breath. "That's nothing."

Ruth shook her head. "It's everything, if it's uneven."

Lena nodded. "Planting schedules slip. Harvest windows stop lining up. Not everywhere—just enough to break predictability."

Paul tapped the table. "Agricultural zones don't just collapse."

"No," Jun said. "They migrate."

Ruth added, "And they don't migrate cleanly."

Silence again.

Paul glanced toward the door, then back. "Climate models would catch that."

Lena's mouth tightened. "They'd *diverge*."

Paul looked up. "Fail?"

"No," she said. "They'd all still run. They'd just stop agreeing."

Jun leaned forward. "Different assumptions amplify differently once predictability decays."

Ruth said quietly, "You get widening envelopes. Longer tails."

Paul shook his head. "That's uncertainty. We already live with uncertainty."

"Yes," Lena said. "But bounded uncertainty."

Paul stared at her. "And this wouldn't be."

"No," Jun replied. "It would be persistent."

Ruth crossed her arms. "And unassignable. No single cause. No single event."

Paul exhaled sharply. "You can't brief that."

Lena nodded. "You can't headline it."

Jun added, "You can't even plot it cleanly."

Paul gave a short, humorless laugh. "So we'd be standing in front of cameras saying what, exactly? 'Everything still works, but less reliably over centuries'?"

Ruth met his eyes. "No one would print it."

Lena looked down at the table. "But we'd still have to live inside it."

No one spoke after that.

These weren't questions for journals.

They weren't questions for policy memos.

They were the kind of questions you asked only when you trusted the room—and even then, barely.

Paul finally pushed his chair back. "This never leaves this room."

Lena didn't answer.

Because they all knew that once you asked *what if* like this, it never really stayed contained.

∝ ∝ ∝

"What could happen" — Without Disaster Sensationalism

Paul was the one who said it, which surprised everyone.

"Insurance breaks first."

Jun looked up. "How?"

"They price risk assuming variance behaves," Paul replied. "Flood tables, drought cycles, storm frequency. You stretch seasonality by weeks, tilt by fractions of a degree, and suddenly the math stops lining up with the payouts."

Ruth nodded. "Not catastrophic failures. Just… bad fits."

Lena added, "Premiums lag reality. Loss ratios creep. Nobody panics. They just adjust coverage."

Paul frowned. "Until they can't."

Jun leaned back. "Infrastructure's worse."

Ruth raised an eyebrow. "Explain."

"Bridges, grids, water systems—everything is built for 'normal' variance plus margin," Jun said. "You don't need extremes to break them. You just need persistence outside the envelope."

Paul nodded slowly. "Brittleness."

"Yes," Jun said. "Quiet brittleness."

Lena tapped the table. "Agriculture wouldn't fail cleanly either."

Paul looked at her. "You're not talking about famine."

"No," she said. "I'm talking about inconsistency. One region misses a planting window. Another gets hit two years later. No pattern you can brief."

Ruth added, "Supply chains hate that."

Jun nodded. "You can't stockpile against drift."

Paul exhaled. "And people move."

Ruth looked at him. "Not refugees."

"Not officially," Paul said. "But pressure increases. Farmers sell early. Coastal towns hedge. Cities absorb quietly until they don't."

Lena frowned. "And there's no single cause."

"No trigger," Jun said. "No disaster date."

Paul rubbed his thumb against the table again, the same nervous habit as before. "Governments would argue endlessly."

Ruth smiled thinly. "About confidence intervals."

"Not outcomes," Lena said.

Jun finished it. "Because nothing *fails* outright."

The room felt tired now. Not alarmed. Tired in the way systems get when they've been compensating too long.

Paul looked around. "You realize none of this sounds like a crisis."

"No," Ruth agreed. "It sounds like noise."

Lena shook her head. "It sounds like slippage."

Paul stared at the darkened screen. "So this doesn't break the world."

"No," Jun said quietly.

"It makes it unreliable."

That word settled harder than any catastrophe ever could.

Paul stood. "That's worse than breaking."

Ruth met his eyes. "Because breaking forces response."

Lena gathered her things. "Slipping gets managed."

No one argued with that.

They left the room knowing something fundamental:

It wasn't disaster.

It was a system slowly losing its grip—and convincing everyone it still had one.

The hallway lights were set to evening mode—dim, amber, designed to ease people out of seriousness. It didn't work.

The door shut behind them with a soft seal. No one spoke for a few steps.

Jun broke first. "So we don't say any of that."

Paul snorted. "We say none of that."

Ruth slowed her pace. "We don't even frame it that way."

Lena glanced back toward the conference room, as if the data might still be listening. "We can't say 'the Moon causes.'"

"No," Paul said. "That's a headline waiting to be misread."

Jun nodded. "We say the assumptions still hold."

Ruth stopped walking. "We say they *should* hold."

Paul turned to her. "There's a difference?"

"Yes," she said. "One is physics. The other is language."

They resumed walking.

Lena exhaled. "So the phrasing becomes conditional."

Jun picked it up immediately. "If current assumptions remain valid—"

"—then axial stability shouldn't vary beyond expected bounds," Ruth finished.

Paul grimaced. "That sounds boring enough to survive review."

"And credible," Lena said. "Because it's true."

Jun frowned. "Until it isn't."

They reached the glass wall overlooking the atrium. People moved below them—normal pace, normal day, unaware of how close language now was to lying.

Paul leaned against the railing. "Here's the problem."

Ruth waited.

"When something *does* happen," Paul continued, "we won't say 'we predicted this.'"

Jun shook his head. "We'll say, 'this wasn't expected under current assumptions.'"

Lena looked at him sharply. "And then those assumptions don't get questioned."

"They get defended," Ruth said quietly.

Paul nodded. "Because no one wants to admit the model was fragile."

Jun swallowed. "So what happens when X shows up?"

No one answered immediately.

Lena finally said, "It shows up barely."

Paul looked at her.

"A week-long seasonal offset," she continued. "A regional mismatch. A forecast divergence small enough to be annoying, not alarming."

Ruth added, "Within error bars."

Jun frowned. "But persistent."

Paul closed his eyes for a moment. "So we keep saying it shouldn't occur."

"And it keeps occurring," Ruth said.

Lena finished it. "Just enough to make the phrase 'should not' start to itch."

They stood there longer than necessary.

Paul straightened. "That's the trick, isn't it."

Jun looked at him. "What trick?"

Paul gestured vaguely, at the building, the institution, the systems built on clean sentences. "We never announce the cause. We announce confidence."

Ruth nodded.

Lena turned toward the elevators. "That way, when X happens—"

"We're still credible," Jun said.

Paul followed them. "And by the time anyone realizes the assumptions are wrong—"

Ruth pressed the call button. "—X is no longer new."

The elevator arrived. Doors opened.

Inside, no one spoke.

Because none of this required catastrophe.

Only time.

∝ ∝ ∝

The Emotional Core

Lena stayed late again.

Leaving felt like abandoning something unfinished, something unnamed. The corridor outside her office had gone quiet. Even the building seemed to be holding its breath.

She pulled up three models on her screen. Same inputs. Slightly different assumptions.

Three answers.

All reasonable.

All defensible.

None agreeing closely enough anymore.

Her phone buzzed. A message from Jun.

Am I crazy, or are these starting to diverge faster than they should?

She stared at it longer than necessary before replying.

You're not crazy.

Three dots appeared. Disappeared. Appeared again.

Then why does it feel like we're the only ones treating this as connected?

She didn't answer right away.

Because that was the isolation—not being alone, but being *unaccompanied.* Everyone else saw pieces. Seasonal drift. Noise. Model variance. Regional anomaly. No one else saw the thread pulling them together.

Across campus, Jun sat at his desk, surrounded by printouts he hadn't meant to generate. He kept telling himself he'd stop once something crossed a threshold.

Nothing ever did.

"It's not wrong enough yet," he muttered, rubbing his eyes.

The outputs still passed review.
The error bars still overlapped.
Every deviation had a plausible excuse.

Which made it worse.

Ruth Calder stood in front of her bathroom mirror, toothbrush forgotten in her hand. She'd spent the evening rewriting the same paragraph in her head—not a paper, not a memo, just a sentence she couldn't make ethical.

If current assumptions remain valid…

They did remain valid. Mostly.

That was the trap.

How do you warn someone about a failure that hasn't failed yet?
How do you sound responsible without sounding speculative?
How do you carry urgency without certainty?

She set the toothbrush down.

This wasn't about brilliance.
It wasn't about discovery.

It was about being early in a way no one thanks you for.

Back in her office, Lena finally typed a reply to Jun.

It's not wrong. It's just not loud.

She stared at the sentence, then added another.

And I don't know how to explain that without sounding like I'm guessing.

Jun read it and felt the familiar, sinking frustration. Not anger. Not fear.

Responsibility.

The kind that comes with seeing the outline of something before it has edges. The kind that doesn't come with instructions.

Across the building, Ruth's phone buzzed. A calendar reminder for a meeting three days out. *Model convergence review.*

She laughed softly. Once.

None of them slept well.

They were holding a truth that wasn't wrong enough to act on—and might never be.

And knowing that didn't feel like insight.

It felt like a burden.

∝ ∝ ∝

Lena began drafting an email she didn't intend to send.

It started as a clarification—three sentences explaining why a minor discrepancy in seasonal alignment had appeared in two unrelated models. She deleted it. Tried again. This time she framed it as a methodological note.

Still wrong.

Anything she wrote sounded either alarmist or trivial. There was no language for *barely consequential but persistently meaningful.* English wasn't built for it.

She saved the draft as *notes_untitled* and closed it.

Jun tried to talk himself out of caring.

He told himself this was normal—models diverged all the time. Assumptions got revised. Papers got corrected. This was how science worked.

Except it didn't feel like revision. It felt like erosion.

He pulled up an old validation run, one he'd trusted for years, and compared it to the current output. The difference was small enough to ignore and large enough to notice.

"Not wrong," he said to the empty room. "Just different."

He hated that phrase. It was how people dismissed things they didn't want to follow.

Ruth sat at her kitchen table long after dinner, the laptop closed with a few pages between the keyboard and the display. Her hands wrapped around a mug that had gone cold. She wasn't thinking about equations anymore. She was thinking about obligation.

At what point does caution become concealment? At what point does restraint become avoidance?

There was no policy for this. No ethical guideline that said, *When confidence drops below X, speak.* Everything assumed clarity would come first.

Clarity wasn't coming.

The next meeting was scheduled. So was the one after that. Reviews, panels, cross-checks—all the mechanisms designed to protect against error were now protecting against movement.

No one was denying the data anymore. That phase had passed quietly.

What they were doing instead was waiting.

Waiting for certainty.

Waiting for the numbers to cross a line bold enough to justify discomfort.

And the numbers refused.

They stayed plausible.

They stayed defensible.

They stayed just inside the boundaries of patience.

Lena caught herself wishing—briefly, guiltily—that something *would* break. That a threshold would be crossed. That a clean error would appear and force a decision.

She hated herself for that thought.

Because wanting disaster, even a small one, meant surrendering something essential.

This wasn't about proof anymore.

The data had already done its job.

What was gone now was ease—the sense that the world's most trusted measurements sat on rock-solid ground, that assumptions could be inherited safely, and that tomorrow would look like yesterday in ways no one needed to question.

The question was no longer *"Is this real?"*

But *"How long do we pretend this isn't actionable?"*

No answers yet.

Just the loss of comfort.

PHASE 3 — INSTITUTIONAL RESPONSE

(How reality is contained without anyone lying)

Nothing was dismissed.

That was the first thing everyone noticed, even if no one said it aloud. The data remained intact. The measurements were preserved. The plots still appeared in shared folders and internal presentations. There were no orders to delete files, no instructions to revise conclusions retroactively. Everything that had been observed was still there, exactly as recorded.

No one falsified data.

The integrity of the work was never in question. Calibration logs remained accessible. Raw returns could still be traced back to their sources. Anyone who wanted to verify the numbers could do so. Transparency, at least in principle, was maintained.

No one suppressed results.

They were not buried or hidden. They continued to circulate through the same channels they always had—reports, reviews, summaries. The information moved forward, but it did so at a different pace, under different conditions. Nothing vanished. It simply took longer to arrive.

No one said the work was wrong.

The science held. That, too, was acknowledged. The models were sound. The measurements defensible. Every technical challenge raised in discussion was addressed carefully, methodically, and in good faith.

The response came instead as process.

Requests for clarification arrived first, framed politely and with professional courtesy. Could this assumption be stated more explicitly? Could that uncertainty be quantified more conservatively? Calls followed for additional validation—independent checks, alternate methods, cross-referenced datasets. Suggestions emerged to widen sample windows, to refine language, to align terminology across teams so that no one could misread emphasis as implication.

Everything was reasonable.

Everything was correct.

Correctness accumulated without conclusion.

Precision multiplied without producing direction.

Agreement formed around method rather than meaning.

The work became harder to challenge and easier to defer.

Momentum shifted from discovery to maintenance.

Each step improved rigor. Each adjustment reduced the risk of overstatement. No single request felt obstructive. In isolation, every change strengthened the work.

The analyses moved forward—just more slowly now.

They were routed through committees whose purpose was caution, whose mandate was stability. Review groups expanded. Drafts circulated through additional layers of feedback. Timelines stretched, thoroughness required time. Meetings were scheduled, rescheduled, postponed, then reconvened with expanded agendas.

Findings were contextualized.

Uncertainties were foregrounded. Confidence intervals moved from footnotes to captions. Conclusions were softened because they were incomplete. Language shifted subtly—verbs hedged, adjectives restrained. Statements that had once been declarative became conditional. Conditional statements became descriptive.

This wasn't denial.

It was classification.

A way of assigning facts to containers that limited their reach.

No one claimed the phenomenon didn't exist. No one argued that the trend was imaginary. The reality of the measurements was accepted throughout the process.

It was containment.

Reality was allowed to exist, but only inside carefully bounded sentences. Only inside charts that emphasized overlap rather than divergence. Only inside phrases that carried procedural weight but little urgency: *insufficient confidence, non-significant deviation, within expected variance.*

These were not evasions. They were terms of art, designed to protect against error. They did their job well.

No one needed to lie.

The system worked exactly as intended.

It prevented premature conclusions. It protected against misinterpretation. It kept speculation from outrunning evidence. Each safeguard operated independently, guided by professional norms and institutional memory.

And in doing so, it gently insulated itself from implications that had no headline, no emergency threshold, no clear moment of arrival.

There was no single decision to delay meaning. No vote to contain it. The process itself absorbed the pressure, dispersing it across workflows, review cycles, and documentation standards. Responsibility fragmented naturally, distributed among roles and procedures designed to manage risk rather than consequence.

The shift from data to analysis had already happened.

Now came the quieter shift—from analysis to management.

Not of the phenomenon.

But of its meaning.

∝ ∝ ∝

The First Principle

The briefing room was smaller than the conference space they were used to. No screens on the walls. Just a long table, a pitcher of water, and a clock that ticked loudly enough to be noticed.

The man from policy introduced himself last. Not by title—by function.

"My role," he said calmly, "is to help translate findings into something usable."

Lena waited for the word *safe*. It didn't come.

He clasped his hands. "Let's start with something important. No one here is talking about suppressing information."

Paul nodded immediately. Jun stayed still.

"We don't hide data," the man continued. "We don't block publication. We don't alter conclusions."

Ruth leaned forward. "Then what do you do?"

He smiled slightly, as if the question had been asked many times. "We classify readiness."

Lena felt something tighten.

He slid a single page across the table. Not a report. A framework.

"Findings move through phases," he said. "Preliminary. Contextualized. Actionable."

Paul scanned it. "And this," he said, tapping the middle column, "is where most things live."

"Yes," the man said easily. "Because most things are complex."

Jun frowned. "Who decides when something becomes actionable?"

"Consensus," the man replied. "Replication. Stability of interpretation."

Ruth's voice was quiet. "And until then?"

"Until then," he said, "it remains incomplete."

Lena spoke before she meant to. "Incomplete doesn't mean incorrect."

"Of course not," he said. "It means not yet usable."

Paul leaned back. "So no one's stopping us from talking."

"No," the man agreed. "You're free to publish, to discuss, to debate."

Jun stared at him. "But nothing changes."

The man met his gaze. "Nothing *official* changes."

Silence settled—not tense, not hostile. Procedural.

"This isn't censorship," the man continued. "It's responsibility. Acting too early creates damage of its own."

For a moment, the room accepted this as reasonable—because it was phrased as care, not control.

Ruth folded her arms. "And acting too late?"

He didn't answer right away.

"That's why we emphasize context," he said finally. "Context slows interpretation. It prevents overreaction."

Lena felt the shape of it then—not a wall, not a lock.

A cushion.

Data wasn't buried.

It was padded.

Wrapped in qualifiers until it could no longer bruise anyone.

Paul nodded slowly. "So the information remains public."

"Yes."

"But classified," Jun said.

"Not classified," the man corrected gently. "Characterized."

Lena looked down at the framework again.

Preliminary.

Context-dependent.

Insufficient confidence to guide response.

Nothing illegal.

Nothing traceable.

Just procedure.

And for the first time, she understood something essential:

Governments didn't need to hide reality.

They only needed to insist it wasn't finished yet.

∝ ∝ ∝

WHO SEES THE DATA FIRST

The first responses didn't come from ministers or press offices.

They came from inboxes that never slept.

Lena noticed the pattern before she named it. Requests arriving from the same kinds of places, written in the same careful tone, asking the same narrow questions.

Not *what does this mean?*

But *can you bound this?*

The space agencies were first. They always were.

A call from orbital analysis. A follow-up from laser ranging. Someone asking whether recent variance could be folded into existing ephemeris updates without revision to navigation tables.

No one sounded alarmed.

"We already expect drift," the man on the line told her. "We just need to know if this changes anything downstream."

Downstream.

Geophysical institutes followed. Long emails, dense attachments, polite requests for clarification on coupling terms—Earth rotation, core-mantle interaction, lunar torque.

Jun forwarded one with a note:

They're not confused. They're narrowing the question.

Climate researchers came next, quieter but more uneasy. Not formal requests—conversations at conferences, side messages, tentative phrasing.

"If tilt variability widened," one of them asked over coffee, "would you expect it to show up as noise or bias?"

Ruth answered honestly. "At first? Noise."

The man nodded, relieved. Noise was manageable.

Defense-adjacent units arrived last, and only indirectly. A meeting that suddenly included navigation specialists. A request framed as a redundancy review. Someone asking whether long-term orbital assumptions were still "robust enough for extrapolation."

No one said *risk*.

They said *resilience*.

Paul noticed what all of them shared.

"They're not surprised," he said afterward. "That's the tell."

Jun nodded. "Because the Moon drifting isn't new."

"It's textbook," Lena said.

Paul frowned. "So what's different?"

Ruth answered without hesitation. "The drift isn't the issue."

Lena continued, "It's what else quietly depends on it staying boring."

Jun added, "Secondary implications."

They all understood now why the response started here.

These groups already lived with slow change.
They already modeled centuries.
They already assumed the Moon moved.

What unsettled them wasn't motion.

It was dependency.

And because they were scientists first, they did what scientists do when something feels off but not broken:
They narrowed scope. They asked for bounds. They kept it technical.

It was the kind of response that felt disciplined—and was therefore trusted.

No escalation.
No politics.
Just a collective instinct to contain novelty by embedding it inside existing frameworks.

Ruth said it aloud later, not accusing—just noticing.

"Everyone who sees this first is trained to treat implication as something you defer."

Paul nodded. "Because that's how you avoid false positives."

Lena looked at the stack of requests on her screen. "And how you miss slow truths."

No one argued.

This wasn't suppression.

It was sequence.

Science first. Meaning later.

And by the time the question reached anyone who dealt in outcomes instead of models, it would already be wrapped in language designed to reassure.

The Moon was still drifting.

That part was old news.

What no one had decided yet—what no one wanted to decide—was whether the things built around that drift were still as stable as everyone assumed.

∝ ∝ ∝

How The Concern Is Framed Internally

The tone shifted without anyone announcing it.

Not in meetings.

In drafts.

Lena saw it first in a shared document—an internal synthesis meant to align terminology across teams. Nothing had been removed. Nothing had been added.

Just adjusted.

"This isn't dangerous," no one wrote.

Instead, the opening line read:

Recent analyses suggest a widening confidence interval in long-horizon projections.

Paul flagged it with a comment. *Accurate.*

Jun added a note beneath it. *Technically precise.*

Ruth didn't comment at all.

In a follow-up call, a geophysicist spoke carefully, as if stepping across ice he trusted but didn't love.

"We may be underweighting a coupling term," he said. "Not incorrectly—just conservatively."

"Which term?" someone asked.

"Lunar torque relative to axial stabilization."

There was a pause.

"Okay," the moderator said. "Let's document that as sensitivity."

Not *risk*.

Sensitivity.

Another meeting. Different agency. Same pattern.

"Long-horizon models are particularly sensitive to small perturbations," a climate analyst said, tapping a slide no one could fault. "This isn't unexpected, but it does merit attention."

"How immediate?" someone asked.

She chose her words carefully. "Not immediate. But cumulative."

Cumulative was safe. Cumulative implied patience.

Paul noticed how often the phrase *should be flagged* appeared.

"We should flag this for future assessment."

"We should flag this for downstream review."

"We should flag this for interdisciplinary context."

Flagging didn't move anything.

It marked it.

Jun leaned over to Lena during one of the calls. "Notice what we're not saying."

She nodded. "We're not saying it matters *now*."

"We're not saying it doesn't," he replied.

Ruth finally spoke up during a smaller session, her voice even, unchallenging.

"If predictability decays before thresholds are crossed," she said, "we may want to consider whether our current framing captures that."

The room went quiet—the sentence was hard to place.

"That's an important nuance," the chair said at last. "Let's note that."

Not *act on it.*

Note it.

Afterward, Paul shook his head slowly. "Do you realize how reasonable all of this sounds?"

Lena did.

Every sentence was true.

Every phrase defensible.

Nothing exaggerated.

Nothing misleading.

And yet, each word took a step backward from consequence.

Not by denial—but by making every implication feel like something that could wait.

"This language isn't deceptive," Jun said later, staring at the final draft. "It's honest."

Ruth nodded. "That's why it works."

Because no one was lying.

They were qualifying.

And qualification, done carefully enough, had a strange effect:

It made concern feel like housekeeping.

It turned implication into maintenance.

It allowed everyone to agree—without agreeing on what to do.

By the time the document circulated widely, it carried no alarm.

Only professionalism.

Only caution.

Only sentences that said exactly what they meant—and nothing more.

And that was enough to disarm urgency before it ever had a chance to speak.

∝ ∝ ∝

The language traveled faster than the concern ever had.

Lena heard it first in a call with orbital analysts who hadn't been in the original loop. Different faces. Same phrases.

"We're seeing a slight widening of the confidence interval," one of them said, scrolling through plots that looked reassuringly dense. "Still well within expected variance."

Paul muted his mic and glanced at Lena. "That's our sentence."

She nodded. Not stolen. Adopted.

At a joint session with space navigation teams, the discussion stayed tightly bounded. Ephemerides were reviewed. Timing residuals examined.

"No need to revise tables," someone concluded. "We'll annotate sensitivity and revisit next cycle."

Next cycle. Six months. A year.

The note went into the record, framed carefully:

Long-horizon orbital solutions remain robust, though minor sensitivity to small perturbations is noted.

Accurate.

Harmless.

Jun joined a NOAA-hosted workshop two weeks later—climate dynamics, seasonal forecasting, nothing overtly lunar. The Moon appeared only in the background assumptions slide, exactly where it had always been.

During Q&A, a modeler raised a hand. "We've noticed some increased divergence in extended projections. Still convergent, but less tightly than before."

Jun felt his stomach tighten.

The panel chair nodded. "That aligns with what we're seeing elsewhere. Likely sensitivity amplification."

Another hand. "Should we be concerned?"

The chair smiled. "Concerned isn't the right word. Aware."

Aware was safe.

Afterward, Jun caught snippets of conversation in the hallway.

"Probably underweighted coupling."

"Long-horizon effects are messy by nature."

"Worth flagging, nothing more."

No one disagreed. No one escalated.

At a geophysical institute, Ruth listened as a senior researcher summarized the issue for a mixed audience.

"This isn't about instability," he said. "It's about predictability margins."

She almost smiled at how clean it sounded.

Someone asked whether this had implications beyond modeling.

"Not directly," he replied. "Any real-world effects would be diffuse and cumulative."

Diffuse. Cumulative. Non-actionable.

The same words. Same order. Same calming effect.

Within months, the phrasing had standardized.

Papers referenced *widening envelopes.*

Briefings emphasized *nonlinear sensitivity.*

Internal notes recommended *continued observation.*

Everyone was speaking the same language now—because it fit everywhere.

Paul pointed it out during a late call. "Do you realize we could hand the same paragraph to NASA, NOAA, or an orbital mechanics group, and no one would blink?"

Lena exhaled. "That's why it spreads."

Because it required no decision—only repetition.

Ruth added quietly, "And why nothing moves."

This was still science talking to science.

Still peers.

Still models, bounds, assumptions.

No politicians yet.

No policy offices.

No press.

Just an expanding circle of experts, all agreeing—correctly—that nothing was wrong enough to act on.

And because the language had already been softened, by the time the concern reached anyone who thought in terms of outcomes instead of equations, it would arrive pre-contained.

Not as a warning, but as context.

The transition hadn't happened yet.

It was coming.

And when government awareness finally arrived, it wouldn't meet raw concern.

It would meet a vocabulary designed to make waiting feel like wisdom.

That was how the handoff would work.

Just smoothly enough that no one could say when responsibility had changed hands.

The strangest part was how quickly it became ordinary.

By the next conference season, the phrases had slipped into slide decks without comment. Not highlighted. Not defended. Just present.

Lena sat in the back row of a packed auditorium, badge flipped backward, listening to a planetary dynamics panel she would have once found dull.

A speaker clicked to a graph showing ensemble spread over long horizons.

"As you can see," he said, "the confidence interval widens slightly beyond standard projections. This is consistent with sensitivity to small perturbations in coupled systems."

No gasp. No questions.

Someone in the audience nodded, satisfied.

During the break, she overheard two researchers comparing notes near the coffee urn.

"Same thing you're seeing?" one asked.

"Yeah. Nothing dramatic. Just… fuzzier edges."

"Probably always was."

"Probably."

Jun had his own version of the moment at a workshop on model validation. A junior analyst presented a poster with unusually wide uncertainty bands.

"Did you change the physics?" Jun asked.

The analyst shook his head. "No. Just stopped constraining one of the coupling terms as tightly."

Jun studied the plot. "And the reviewers?"

"They liked the honesty," the analyst said. "Said it was more realistic."

More realistic.

That word followed Jun for days.

Ruth noticed the shift in tone during peer review. Comments that once would have demanded tighter convergence now praised caution.

Appropriate restraint given long-horizon uncertainty.

Conclusions carefully bounded.

Speculation appropriately minimized.

No one was pushing back anymore. They were settling in.

At a joint session between climate and orbital groups, a moderator summed it up with professional calm.

"We're converging on a shared understanding," she said. "These systems are sensitive, nonlinear, and best handled with humility."

Heads nodded. Applause followed.

Afterward, Paul leaned against a wall and laughed softly. "Do you hear that?"

"Hear what?" Lena asked.

"That sound," he said. "That's the community agreeing to live with it."

"With what?" Jun asked.

Paul gestured vaguely. "With not knowing cleanly anymore."

They didn't mean ignorance.

They meant managed uncertainty.

Months passed. Papers were published. Grants renewed. Models updated.

The Moon drifted, just as it always had.
The equations still closed.
The forecasts still ran.

But something subtle had changed: the expectation of convergence.

People stopped waiting for the lines to come back together.

They accepted spread as the cost of realism.

Once that trade was accepted, there was no longer any pressure for the lines to come back together.

At that point, uncertainty stopped feeling provisional and began feeling structural.

Ruth said it out loud once, late, after a long day.

"We've crossed a line."

Lena looked up. "Which one?"

"The one where uncertainty is no longer a problem to solve," Ruth said. "It's a condition to manage."

Jun frowned. "That's not wrong."

"No," Ruth agreed. "But it changes who feels responsible."

Because once uncertainty became normal, urgency had nowhere to stand.

The scientific community hadn't ignored the issue.
They hadn't denied it or even downplayed it.

They had absorbed it.

And by doing so, they made it safe enough to pass along.

The next step wouldn't come from science.

It would come from the moment someone outside the models asked a different question—not *is this true?*, but *what do we do with it?*

That moment hadn't arrived yet.

But the ground was already prepared.

∝ ∝ ∝

THE FIRST CROSS-BOUNDARY MEETING

The room wasn't intimidating.

That was the unsettling part.

That absence was deliberate.

Comfort had been engineered into the room.

It looked like every interagency space Lena had ever been in—neutral walls, a modular table, lighting tuned for alertness without strain. No flags. No insignia. Just name cards printed in the same font.

Observer.

Liaison.

Coordinator.

Not titles. Functions.

The screen displayed a familiar graph. The same one they'd been carrying from room to room for months, stripped now of anything that might suggest emphasis.

"This won't take long," the coordinator said pleasantly. "We're here to align understanding."

Align. Not decide.

Lena spoke first, because she always did. Silence created room for misinterpretation.

"The observed divergence remains within expected bounds," she said. "However, sensitivity increases over longer horizons."

The liaison nodded. "So this is a modeling issue."

"It's a modeling *property*," Ruth corrected gently. "The models are behaving as designed."

"Good," the liaison said. "That's important."

Jun shifted in his chair.

Paul watched the room carefully. This wasn't skepticism. This wasn't pushback.

This was relief.

Relief meant there would be no pressure to decide anything today.

A man from strategic assessment leaned forward. "To be clear, no one here is suggesting immediate risk?"

"No," Lena said. "Nothing immediate."

"And no thresholds have been crossed."

"Correct."

"So if current assumptions remain valid—"

Lena felt the phrase settle in her chest before she spoke it.

"—then we shouldn't expect actionable deviation."

The coordinator wrote it down.

Just like that, the sentence changed hands.

Ownership had shifted without anyone needing to agree on what it meant.

Discussion followed, but it circled inward. Clarifications. Boundaries. Timelines that extended conveniently beyond decision cycles.

"How often should this be revisited?" someone asked.

"Annually," Ruth said. "Or as part of routine review."

"Excellent."

Paul finally spoke. "Just to be explicit—this isn't being escalated?"

The coordinator smiled. "Escalation implies urgency. We're talking about stewardship."

Stewardship sounded noble.

The meeting ended on schedule. No tension. No raised voices.

As they packed up, Jun whispered, "They didn't ask what happens if the assumptions don't hold."

Paul replied quietly, "They don't have to."

Outside the room, Lena paused.

"That was the handoff," she said.

Jun frowned. "It felt like nothing."

Paul shook his head. "That's how you know it worked."

∝ ∝ ∝

THE SENTENCE THAT GETS CORRECTED

Jun didn't think of himself as rebellious.

He liked clean math. Careful phrasing. Review processes that caught mistakes before they metastasized.

But the paragraph bothered him.

He'd been asked to draft a section on long-horizon implications — not outcomes, just behavior. He wrote carefully, deleting anything that sounded speculative.

Still, one sentence refused to soften.

If predictability continues to decay, secondary systems may experience structural misalignment prior to threshold events.

He reread it six times.

It was true.

He submitted the draft.

The feedback arrived the next morning.

Polite. Fast. Anonymous in tone.

This phrasing is technically correct but may be interpreted as outcome-oriented. We recommend reframing to focus on model characteristics rather than downstream effects.

A suggested edit followed, immaculate in its restraint:

Long-horizon models exhibit increased sensitivity to small perturbations, warranting continued observation.

Jun leaned back.

Nothing had been removed.

Nothing contradicted.

The revised sentence was safer because it pointed nowhere.

Direction, once removed, was very difficult to reintroduce.

He typed a reply: *Understood. Revised.*

Later that day, Ruth stopped by his office.

"They flagged your sentence," she said, not accusing.

"I know."

She hesitated. "You weren't wrong."

"I know."

She nodded. "That's the problem."

Jun looked at her. "Do you think I should have pushed?"

Ruth thought for a long moment. "Pushed toward what?"

He didn't answer.

Because there was no direction yet. Only drift.

∝ ∝ ∝

The Internal Memo

The memo didn't arrive all at once.

It began as a request.

Lena received it first — a calendar invite titled *Synthesis Alignment Session.* No agenda. No attachments. Just a list of attendees that was longer than it needed to be.

She stared at the names longer than she should have. Too many intermediaries. Too many people whose job descriptions involved verbs like *coordinate*, *integrate*, *translate*.

When the meeting started, no one mentioned urgency.

The chair opened with a sentence that would repeat, in one form or another, for the next six months.

"We're not here to change the science," he said. "We're here to make sure it's being represented consistently."

Consistently was the word that mattered.

It implied alignment without judgment.

It framed interpretation as a coordination problem.

Draft One appeared a week later.

It was rough in the way institutional writing always is at first — too many voices, too many caveats stacked awkwardly on top of one another. But the spine was already visible.

No conclusions yet.

No recommendations.

Just framing.

Jun read it on a train ride home and felt an unexpected flicker of relief.

Nothing was wrong.

Every sentence tracked. Every claim was defensible. If he'd been handed this cold, he would have nodded along.

Still, he opened a document and began annotating.

"Confidence interval widening" — accurate, but specify horizon. "Nonlinear sensitivity" — maybe cite ensemble behavior. "Secondary implications" — avoid outcome language.

He didn't notice when he stopped pushing back and started refining.

Draft Two folded in feedback from space agencies.

Navigation teams wanted reassurance that ephemerides remained intact.

Laser ranging groups emphasized continuity of measurement.

Orbital dynamics specialists requested a paragraph clarifying that no thresholds had been crossed.

All of it reasonable.

Lena noticed how often the word *robust* appeared now.

Robust models.

Robust frameworks.

Robust assumptions — *within defined parameters.*

She suggested a footnote clarifying what those parameters were.

The footnote survived. Barely.

Draft Three incorporated climate perspectives.

NOAA's contribution was careful, almost artful. They didn't argue with the science. They reframed it.

Extended projections show increased spread.

Spread does not imply failure.

Ensemble divergence is expected in complex systems.

Expected was doing heavy lifting now.

Ruth read that section three times and finally said, out loud, to no one, "Expected by whom?"

But she didn't write that down.

By Draft Four, the memo had a tone.

Not defensive.

Not dismissive.

Mature.

This was how senior science spoke when it wanted to be taken seriously by people who didn't think in equations.

The language had slowed.

Sentences lengthened.

Verbs softened.

Causality dissolved into correlation, then into description.

Jun caught himself deleting a phrase — *prior to threshold events* — and replacing it with *absent discrete triggers.*

He paused.

The meaning hadn't changed.

The weight had.

By the fifth draft, contributors stopped tracking changes.

They trusted the document now.

That trust was dangerous.

The memo circulated "for awareness" across departments that had not been in the loop. No questions came back. No objections.

That was its own kind of feedback.

Paul noticed something else.

"Look at the order," he said during a late call.

"Order of what?" Lena asked.

"The structure. Data first. Methods second. Interpretation last."

"That's normal."

"Yes," Paul said. "But consequences aren't anywhere."

Jun frowned. "They're implicit."

Paul nodded. "Exactly."

Draft Seven added an executive summary.

Three paragraphs. No bullet points.

It could be read aloud without raising a voice.

The final sentence read:

Continued monitoring and interdisciplinary coordination are recommended to ensure long-horizon resilience.

Resilience had entered the document quietly and refused to leave.

When the memo was finalized, it was forty-two pages long.

Dense. Polite. Impeccably sourced.

It was sent upward with no cover letter.

It didn't need one.

Jun's reward arrived two days later, and it didn't feel like a reward at first.

An invitation to join a standing working group. Expanded access. Travel authorization. A note praising his *measured contributions.*

Measured.

He read the word several times.

Measured meant trustworthy.

Measured meant promotable.

Measured meant safe.

He accepted.

He wanted to stay inside the room where decisions *might someday* be made.

The memo began to appear everywhere.

Cited in briefings.

Referenced in grant proposals.

Used as baseline context.

It became the thing people pointed to when they didn't want to re-open the question.

"This is already covered," someone would say.

And it was.

Perfectly.

The strangest part was that Jun still believed every word of it.

That was what haunted him later.

Not that they'd lied.

Not that they'd hidden anything.

But that they had produced a document so accurate, so careful, so calm — that it made action feel unnecessary without ever saying so.

The memo didn't end the conversation.

It made sure the conversation would never escalate on its own.

And that, somewhere around page seventy-nine, was when comfort finally left for good.

∝ ∝ ∝

The Memo Becomes the World

The first deviation arrived disguised as housekeeping.

It came through NOAA, buried in a routine seasonal outlook update that barely warranted a meeting. The email subject line was apologetic in tone, almost embarrassed by its own existence.

Minor calibration adjustment — long-range ensembles.

Jun almost didn't open it.

When he did, he felt the now-familiar tightening — not alarm, not excitement, just recognition.

The ensembles still converged. Mostly. But the convergence took longer. The overlap was thinner. The tails stretched a little farther than before.

Nothing broke.

He forwarded it to Lena with a single line.

This fits.

She replied an hour later.

Route it through the memo.

That was the first time either of them had said it like that.

Not *compare it.*

Not *analyze it.*

Route it.

The memo had become a verb.

It no longer described reality; it routed it.

The follow-up meeting was brief.

A NOAA modeler walked through the update carefully, making sure not to lean on any word that could be misheard.

"We're seeing slight seasonal phase drift," she said. "Weeks, not months. Regionally inconsistent."

Paul nodded. "Within expected variance?"

"Yes," she said. "If we apply the long-horizon sensitivity framing."

There it was.

No one challenged her.

No one needed to.

The memo already provided the language that made the deviation safe.

"Then we cite the synthesis," the chair concluded. "Note alignment."

Alignment again.

Jun scribbled something in the margin of his notebook and then crossed it out.

First X.

He didn't write it down.

Two months later, infrastructure analysts flagged a discrepancy in freeze-thaw modeling — nothing dramatic, just a mismatch between historical assumptions and updated projections.

The question wasn't *why.*

It was *where to file it.*

"This falls under long-horizon sensitivity," someone said during the call.

"Agreed."

The memo appeared on the shared screen, not as evidence, but as permission.

Once permission replaced judgment, deviation no longer needed to be understood.

By now, everyone knew which paragraph to quote.

Not the one about uncertainty.

The one about *resilience through monitoring.*

Ruth noticed it during peer review.

A paper came back with unusually gentle comments.

Results are interesting but consistent with previously characterized model sensitivity.

No revision necessary at this time.

She stared at the phrase *at this time.*

Time had become elastic.

The deviation count increased, slowly enough that no one tracked it as a set.

A planting window shifted here.

A forecast lagged there.

A regional mismatch that resolved itself just enough to avoid attention.

Each one was small.

Each one was real.

Each one was routed through the memo and neutralized without anyone ever saying *neutralized.*

Lena caught the pattern one evening while scrolling through updates she hadn't meant to review all at once.

"These aren't independent," she said quietly.

Jun looked up. "They don't have to be."

"That's not what I mean."

She turned the screen toward him. "They're all being interpreted through the same document."

He didn't argue.

Because that was the point.

The memo wasn't just context anymore.

It was the lens.

Anything that fit inside it became manageable. Anything that didn't… didn't arrive.

That realization came later.

Not as a revelation — as a pause.

Jun was preparing a response to a new inquiry when he noticed something odd. The question being asked wasn't the one the data raised.

It was narrower.

Does this deviation exceed the bounds described in the synthesis memo?

Not *what does this suggest?*

Not *is this changing?*

Just: does it exceed the memo?

He leaned back in his chair.

At some point, the memo had stopped being a summary of understanding.

It had become the definition of relevance.

Anything that didn't challenge it directly ceased to exist as a problem.

He brought it up with Paul during a late call.

"We're no longer asking what the data means," Jun said. "We're asking whether it contradicts the memo."

Paul was quiet for a long time.

"That's how baselines work," he said finally.

"That's how they harden," Jun replied.

Paul didn't disagree.

The next deviation was larger — still not dramatic, but harder to wave away. A multi-region climate model showed persistent divergence that didn't collapse under recalibration.

NOAA flagged it internally. Carefully.

The discussion took longer this time.

"Still within expected bounds," someone said.

"Barely," someone else replied.

Silence.

Then: "Does it fall under previously documented sensitivity?"

The memo came up again.

It always did.

The conclusion was the same as before — *monitor, reassess next cycle* — but something had shifted in the room.

Awareness.

Lena felt it walking back to her office.

The memo had done its job too well.

It had made everything survivable.

It had also made escalation structurally impossible.

Later that night, Jun reopened his original draft — the one with the sentence that hadn't survived.

He read it slowly.

It wasn't prophetic.

It wasn't dramatic.

It just pointed *forward.*

He understood then what he'd been rewarded for.

Not being careful.

Being containable.

The memo had become the final word—because it decided which questions were allowed to exist.

Outside, the world still looked stable.

Inside the systems that measured it, something had begun to slip — not breaking, not failing.

Just drifting far enough that comfort could no longer return.

And no one, not even the people closest to the data, could say exactly when the moment to act had passed.

Only that it felt like it already had.

The First Thing That Doesn't Fit

It wasn't larger than the others.

That was the mistake.

If it had been dramatic—if a model had blown up, if a forecast had inverted—there would have been language for it. Protocol. Escalation paths already rehearsed.

This wasn't that.

It came in as a reconciliation failure.

Two long-horizon models, built by different teams, using different architectures, refused to converge even after calibration. Not by much. Not visibly. But enough that the usual smoothing tricks didn't work.

Jun stared at the overlay until his eyes hurt.

"You see it," he said finally.

Lena leaned closer. "I see it."

Ruth, on the speakerphone, said nothing at first.

"It doesn't violate the memo," Jun continued. "But it doesn't sit inside it either."

Paul exhaled. "Say that again."

"The memo assumes widening confidence," Jun said. "This isn't widening. It's *splitting*."

That distinction mattered because the memo had only been built to absorb blur, not divergence.

Silence.

Ruth spoke carefully. "Does it resolve if you shorten the horizon?"

"Yes," Jun said. "If you make it someone else's problem."

That landed.

Lena scrolled. "If this were just noise, the ensembles would blur. They're not. They're choosing different futures."

Paul closed his eyes for a moment. "That sentence cannot go in writing."

"I know," Jun said. "That's the problem."

They tried to route it through the memo anyway.

Of course they did.

Jun drafted a framing note that used every approved phrase he could find.

Increased sensitivity.
Persistent divergence.
Within extended uncertainty bounds.

He sent it out.

The response came back an hour later, polite and unusable.

This appears consistent with previously characterized long-horizon behavior.

Jun stared at the message.

It wasn't wrong.

It was incomplete in a way the memo could no longer cover.

Lena called the NOAA contact directly, bypassing the chain.

"Have you seen this pattern before?" she asked.

A pause. Paper shuffling.

"No," the woman said finally. "But it doesn't exceed anything we've agreed is actionable."

"Agreed where?"

Another pause. Shorter this time.

"In the synthesis."

After the call, Lena sat back in her chair.

"That's it," she said.

Paul frowned. "That's what?"

"The memo is now upstream of observation."

Ruth added quietly, "Which means science can't move this forward anymore."

It no longer had a place to put the result.

They didn't argue.

Because the problem wasn't that the deviation contradicted the memo.

It was that it **required a new one**.

Jun tried anyway.

He drafted an addendum.

Not a revision. Not a challenge. Just a note: *This behavior is qualitatively different.*

The response was swift.

We recommend folding this into existing language to maintain continuity.

Continuity.

Jun deleted the draft.

Later, alone, he pulled up the plots again and tried to explain them to himself without using any of the approved words.

He failed.

The vocabulary had been exhausted.

The next meeting included people who hadn't been there before.

Not scientists.

Listeners.

They didn't ask technical questions. They asked procedural ones.

"How confident are you that this falls outside prior characterization?"

"What would it take to say that definitively?"

"Is there a risk of overinterpreting early divergence?"

Lena answered carefully.

Ruth answered conservatively.

Jun answered less than he wanted to.

No one pushed.

No one challenged the science.

But something had shifted in the room: the center of gravity.

This was no longer about whether the models were right.

It was about whether the uncertainty could still be *owned* by the people who found it.

Afterward, Paul said it out loud for the first time.

"This is no longer a scientific containment problem."

Lena nodded. "Because it doesn't fit the container."

Ruth finished the thought. "Which means it goes somewhere else."

Jun looked up. "Where?"

No one answered immediately.

They didn't need to.

Because when something stops fitting inside science, it doesn't disappear.

It escalates by default.

To awareness.

The next email didn't come from a research institute.

It came from an office that didn't use equations in its subject lines.

The phrasing was careful. Neutral. Almost friendly.

We've been briefed on recent discussions regarding long-horizon uncertainty. We'd appreciate an opportunity to understand any potential cross-domain implications.

Jun read it twice.

"This is it," he said.

Lena closed her laptop. "Yes."

Paul nodded slowly. "This is when it changes."

The data could no longer be described using language that belonged only to science.

They didn't say anything else.

The door they'd been leaning against for months—carefully, professionally—had finally opened on its own.

And on the other side wasn't fear.

It was governance.

∝ ∝ ∝

When Governments Become Aware

Governments didn't enter the conversation because the science changed.

They entered because spreadsheets stopped agreeing with each other.

Agreement, not accuracy, was the first thing to break.

The first signal came from a budget office that had no interest in the Moon.

An analyst flagged a discrepancy during a routine reconciliation — long-term infrastructure allocations drifting just enough to complicate out-year projections. Nothing alarming. Just misalignment.

Seasonal assumptions used by one agency no longer matched those used by another.

The note was forwarded with a single question attached:

Which model are we using?

No one answered right away.

The second signal arrived through defense, quietly and sideways. A request for clarification routed through a research liaison, phrased so carefully it barely looked like a request at all.

"Some long-horizon navigation and stability assumptions appear sensitive to recent updates," the message read. "We're seeking confirmation that existing baselines remain appropriate for planning purposes."

Appropriate was the word doing the work.

Not *accurate.*

Not *correct.*

Appropriate for planning.

Lena read it twice before forwarding it to the group.

"They're not asking what's changing," Jun said. "They're asking whether they can keep using what they already published."

"That's always the first question," Paul replied.

Climate policy followed because timelines had begun to slip out of sync.

Targets based on seasonal predictability stopped aligning cleanly across regions. Policy drafts that once nested neatly inside model ranges now required footnotes.

Too many footnotes.

A senior advisor asked during a briefing, "Why are our confidence bands wider than they were last year?"

The answer was honest.

"Improved realism."

The advisor nodded, satisfied.

But later, a different question surfaced — not in the room, not on the record.

Does this affect commitments we've already made?

That was the trigger.

Not fear or curiosity.

Exposure.

By the time a formal interagency session was convened, the language had already been set by science itself. The memo circulated ahead of time, highlighted, excerpted, stripped of its nuance and presented as a reference document.

No one challenged it.

They relied on it.

A deputy secretary opened the meeting with practiced efficiency.

"We're not here to revisit fundamentals," she said. "We're here to understand whether existing stability assumptions remain valid for planning."

Valid.

Not *true*.

Not *sufficient*.

Valid within previously published frameworks.

Validity meant compatibility with what already existed, not adequacy for what was emerging.

Jun felt the shift immediately. The room wasn't hostile. It wasn't skeptical.

It was protective.

"What we need," the deputy continued, "is confidence that nothing here undermines long term projections already in use."

Ruth spoke carefully. "The science doesn't contradict them."

That sentence had become reflex.

"But," she added, "it does suggest increased sensitivity over extended horizons."

The deputy nodded. "Sensitivity that we have already characterized?"

"Yes," Lena said. "Within the bounds described in the synthesis."

That was all anyone needed.

A policy advisor leaned forward. "So this doesn't require re-baselining existing commitments."

"No," Paul said. "Not at this time."

At this time passed through the room without resistance.

The deputy smiled. "Good. Then our focus should remain on monitoring and coordination."

Monitoring was safe.

Coordination was controllable.

No one asked what would happen if the assumptions stopped holding. That question belonged to science — and science had already given them a way not to ask it.

After the meeting, Jun stood in the hallway longer than necessary.

"They didn't come because of the data," he said.

Lena nodded. "They came because of alignment."

Paul added, "Because budgets, infrastructure, defense, and policy all assume the same future."

"And they needed to know," Ruth finished, "whether that future was still safe to reference."

No one said *wrong*.

No one said *unstable*.

The concern had entered government exactly the way it always did — not as discovery, but as **risk to prior statements**.

From here on, the question wouldn't be whether the science was sound.

It would be whether acknowledging its implications would force revisions that no one wanted to make.

The machinery had engaged.

Not to respond.

To preserve consistency.

And once that happens, truth doesn't disappear.

It just has to wait its turn.

∝ ∝ ∝

The next meetings were quieter.

Not smaller — quieter.

Attendance lists stabilized. The same names appeared again and again, suggesting consensus where there was really just habit. Slides were reused. Language hardened.

A senior planner opened one session with a statement that sounded reassuring until it was examined closely.

"We're operating under the assumption that published stability baselines remain serviceable."

Serviceable.

Jun wrote it down and underlined it twice.

The discussion that followed never touched the Moon.

It didn't need to.

Instead, it orbited secondary effects — how long-range infrastructure amortization models handled widened uncertainty bands, how defense planning windows tolerated drift so long as it stayed symmetrical, how climate policy could absorb variance without reopening targets already ratified.

"None of this invalidates our frameworks," a budget officer said. "It just stretches them."

Stretching was acceptable. Breaking wasn't.

A defense analyst spoke next, careful and measured.

"Our navigation models assume axial stability within a defined envelope," she said. "We're confirming that this envelope still holds under updated sensitivity analyses."

Lena answered truthfully. "It holds."

"What about edge behavior?" the analyst asked.

Lena paused. "Edge behavior is less predictable."

The analyst nodded. "But still bounded."

"Yes," Ruth said. "For now."

For now passed without comment.

Later, during a smaller breakout, a climate policy advisor leaned toward Paul.

"You're not telling us something will fail," he said.

"No," Paul replied.

"But you're also not telling us it won't."

Paul hesitated. "We're telling you the models are behaving differently under the same assumptions."

The advisor considered that. "Different enough to matter politically?"

Paul didn't answer immediately.

Jun did. "Not yet."

The advisor smiled, relieved. "Then we're aligned."

Alignment again.

Afterward, Jun walked with Lena down a long corridor that smelled faintly of cleaning solvent and old paper.

"They're asking the wrong question," he said.

"They're asking the only question they can," she replied.

"Which is?"

"Whether acknowledging this forces them to change what they've already said."

Jun stopped walking. "And if it does?"

Lena looked straight ahead. "Then they'll need certainty they don't have."

That evening, a briefing packet circulated upward — not alarming, not detailed. Just enough to ensure senior leadership wasn't surprised later.

The executive summary opened with a sentence Jun recognized immediately.

Current analyses don't indicate a need to revise existing stability assumptions.

He closed his eyes.

The second paragraph followed.

However, continued monitoring is recommended as long-horizon sensitivity remains an active area of research.

Active was the word now.

Not concerning.

Not urgent.

Active.

The packet was acknowledged. Logged. Filed.

No directives issued.

But something subtle happened after that.

Questions stopped flowing *downward.*

Instead, requests began to flow *sideways* — coordination calls, alignment checks, phrasing reviews.

Language audits.

A policy aide emailed Lena late one night.

When discussing divergence, should we continue to reference "model behavior" rather than "system behavior"?

Lena stared at the question longer than she should have.

She replied carefully.

Yes. That remains the most accurate framing.

She knew what she was doing as she typed it.

She wasn't lying.

She was preserving coherence.

Weeks passed.

Government awareness deepened, but it didn't widen. The issue didn't move up the agenda — it settled into a category.

Long-horizon uncertainty.

That category had a shelf life measured in years.

Jun realized it during a meeting when someone said, almost casually, "This isn't a near-term stability issue."

No one corrected him.

Near-term had become the only timescale that mattered.

Anything beyond that horizon could be acknowledged indefinitely without ever needing to be resolved.

After the meeting, Ruth spoke quietly.

"We're past the point where science leads."

Paul nodded. "Now science supports."

Lena finished the thought. "And language governs."

They all felt it then: the loss of another kind of comfort.

Not the comfort of certainty — that had gone earlier.

The comfort of relevance.

Because once governments become aware in this way—cautiously, procedurally, protectively—the question is no longer whether something is real.

It's whether reality can be acknowledged without destabilizing the structures already built on denying its urgency.

And until that answer exists, awareness doesn't lead to action.

It leads to management.

Containment had crossed its final boundary.

From here on, the system would do what systems do best:

Optimize for continuity.

Even if continuity slowly drifted away from truth.

The task force wasn't called a task force.

It was a *working alignment group*, chartered for six months with an option to renew. Its mandate was modest: ensure consistent interpretation of long-horizon uncertainty across domains.

No one objected to that.

The first meeting agenda was three pages long and contained nothing that looked like a decision point. Updates. Crosswalks. Terminology review. A breakout on "external messaging considerations," clearly marked *internal only*.

Jun noticed how quickly people stopped asking who had initiated it.

Initiation implied agency.

This felt more like gravity.

A policy analyst opened with a familiar reassurance. "We're not here to change baselines."

Heads nodded.

"We're here to make sure everyone understands how to talk about them."

That phrasing stuck.

The discussion that followed was almost entirely about adjectives.

Was *widening* preferable to *expanding*?

Did *sensitivity* imply fragility, or just responsiveness?

Was *nonlinear* too technical for mixed audiences?

Lena listened, answering when asked, careful to stay inside the lane she'd helped pave.

When someone suggested replacing *predictability decay* with *forecast dispersion*, she hesitated.

"That's not wrong," she said finally. "But it shifts emphasis."

"To where?" the chair asked.

"To outputs," Lena replied. "Not causes."

The chair smiled. "That's intentional."

After the meeting, Ruth caught up with her near the elevators.

"We're editing the future," Ruth said quietly.

"No," Lena replied. "We're editing how it can be referenced."

Ruth didn't argue.

Over the following weeks, the group's influence spread laterally, moving through departments and agencies that rarely coordinated directly but were now sharing phrasing guides. Slide templates appeared with pre-approved language blocks. Briefings began to sound eerily similar, regardless of who delivered them.

Consistency became a virtue.

Jun sat through one presentation where a senior official summarized months of work in two sentences:

"We're seeing increased long-horizon sensitivity in some coupled systems. This does not presently affect published stability assumptions."

Applause followed. Polite. Appreciative.

Jun felt something hollow out behind his ribs.

Later, alone, he opened the raw plots again. The split he'd seen months earlier was still there. It hadn't grown dramatically. It hadn't resolved.

It had persisted.

He brought it up during a smaller session, choosing his words carefully.

"This behavior remains unresolved," he said. "It doesn't map cleanly onto our existing characterization."

The response came gently.

"Unresolved doesn't mean unmanageable," someone said.

Another voice added, "And unmanageable doesn't mean actionable."

Jun nodded, because that was true too.

That was always the problem.

A new phrase entered circulation shortly after: *decision-relevant horizon.*

It appeared first in a budget memo, then in defense planning notes, then in climate policy briefs.

Anything beyond that horizon was acknowledged, documented, and set aside.

Not denied.

Deferred.

Lena noticed how often people invoked it to close conversations.

"That's outside the decision-relevant horizon."

"We'll need more data before it becomes decision-relevant."

"This doesn't cross into decision relevance."

The horizon moved as needed.

Government awareness, she realized, didn't mean comprehension.

It meant categorization.

By the time senior leadership referenced the issue publicly—once, obliquely—it had already been fully digested.

A spokesperson answered a question about long-term climate uncertainty with practiced calm.

"Models are continually improving," he said. "Some spread is expected as realism increases. There is no indication this affects near-term stability."

The sentence was technically perfect.

It was also the end of that line of questioning.

Inside the system, work continued. Papers were published. Models refined. Deviations logged.

Outside, nothing appeared to change.

Jun met Paul for coffee one morning, both of them pretending it was casual.

"We're past the point of warning," Jun said.

Paul stirred his cup slowly. "We were never at a point of warning."

"Then what were we at?"

Paul thought for a long moment. "Translation."

Jun nodded. "And now?"

"Now," Paul said, "we're at maintenance."

That word lingered between them.

Maintenance implied care.

Maintenance implied stewardship.

Maintenance also implied keeping something running as designed—even if what it was running toward no longer matched the original intent.

Weeks later, Ruth declined an invitation to speak at a conference. Not out of fear. Out of fatigue.

"They want reassurance," she said to Lena. "And I don't know how to give it without lying."

"You wouldn't be lying," Lena replied.

Ruth looked at her. "That's worse."

By the time the working group's charter came up for renewal, no one questioned it.

It renewed automatically.

Government awareness had matured into something quieter than concern and heavier than ignorance.

Ownership.

Ownership of the framing, not the consequences.

From here on, every new data point would be interpreted through structures already optimized for continuity. Every deviation would be assessed not for what it suggested, but for what it required.

And what it required—again and again—was patience.

The system had absorbed the truth as far as it could without changing shape.

What came next wouldn't be decided by discovery.

It would be decided by how long that shape could hold.

And for the first time, Lena understood the real danger:

Not that the world might change.

But that the institutions built to notice change had already learned how not to feel it.

∝ ∝ ∝

The Unspoken Economic Problem

It didn't surface in an emergency briefing. It resurfaced in a revision request.

The kind that arrives quietly, without urgency or emphasis. No alarms accompanied it. No one treated it as time-sensitive.

A senior economist flagged a discrepancy during a routine, long-term, outlook update — nothing dramatic, just a mismatch between growth projections and infrastructure depreciation curves that used to align cleanly.

She wrote one sentence in the margin:

Assumptions appear increasingly sensitive to seasonal variance.

The sentence wasn't alarming—just expensive.

No one replied to that sentence directly.

They scheduled a call.

The call opened with reassurance.

"We're not revisiting macro fundamentals," the chair said. "This is about parameter consistency."

Consistency again.

The economist shared her screen. Rows of numbers. Columns stretching decades forward. The kind of models that never made headlines but decided everything underneath them.

"These projections assume climate variability stays within historically bounded envelopes," she said. "Those envelopes are widening."

A budget analyst frowned. "Widening doesn't mean invalid."

"No," she agreed. "It means pricing gets harder."

Someone asked the obvious question without meaning to.

"Harder for whom?"

She didn't answer immediately.

"For insurance," she said finally. "For municipal bonds. For agricultural lending. For long-horizon capital."

Silence — not alarmed, just attentive.

"These instruments don't fail when risk increases," she continued. "They fail when risk stops being classifiable."

Unclassifiable meant unpriceable, and unpriceable meant nobody could sign their name.

A defense finance liaison spoke next, cautious.

"Our procurement models assume stable amortization schedules. Variance is acceptable. Drift is manageable. But unpredictability—"

He stopped himself.

"But nothing here suggests unpredictability," someone said quickly.

The economist didn't argue.

She adjusted the slide.

"What I'm saying," she said carefully, "is that the cost of hedging uncertainty is rising faster than our models assume."

That was safer.

Costs could be managed.

Later, in a smaller meeting, the language sharpened slightly.

"If planting windows shift inconsistently," someone said, "credit risk becomes regional rather than sectoral."

"Which breaks pooling," another replied.

"And if pooling breaks," the economist added, "premiums rise unevenly."

No one said *instability.*

They said *pricing distortion.*

The memo appeared again — not as authority, but as boundary.

"Does this contradict the synthesis?" someone asked.

"No," came the answer. "It's consistent with long-horizon sensitivity."

Then we note it, the chair said.

Noting was how the system admitted reality without accepting responsibility for it.

They noted it.

But the economist kept going.

"What worries me," she said, "is that none of this triggers a correction. It just accumulates friction."

Friction was a word people understood.

"Growth slows," she continued. "Not everywhere. Not dramatically. Just enough that long-term projections require more caveats."

Caveats were manageable.

"What about public confidence?" someone asked.

She shook her head. "This won't show up that way."

"How will it show up?"

She thought for a moment. "As higher costs justified by prudence."

That sounded almost reassuring.

After the meeting, she sent a follow-up note to a small list.

This does not constitute economic risk in the traditional sense. However, it may introduce persistent inefficiencies across long-horizon planning.

Persistent inefficiencies.

No headlines.
No alarms.

Just a quiet tax on predictability.

And as long as the answer fit inside a spreadsheet, the assumptions could remain sacred.

Jun read the note later that evening, forwarded from someone he barely knew.

He understood immediately why this part was never said out loud.

You could argue with science.
You could debate models.
You could defer uncertainty.

But once the numbers touched money — not markets, but commitments — the problem stopped being abstract.

It was cumulative.

And cumulative costs didn't trigger action.

They changed behavior.

Slowly. Quietly. Everywhere.

No one announced it.

No one needed to.

The unspoken economic problem wasn't that the world might become unstable.

It was that stability had been quietly priced into everything—and no one knew how to update the price without admitting that the assumptions behind it were no longer solid.

Still inside government.

Still no reaction.

Just the first time someone noticed that continuity itself had begun to carry a cost.

The problem didn't escalate because it didn't violate any rule.

Every model still ran.
Every forecast still produced a number.
Every budget still balanced—on paper.

What changed was how often qualifiers appeared.

A Treasury briefing flagged *increased variance assumptions* in long-range revenue models. Not enough to revise projections, just enough to require a footnote explaining why confidence bands had widened.

The footnote was approved without discussion.

At a transportation planning session, an analyst noted that infrastructure lifetimes were being stress-tested against a broader range of seasonal conditions than before.

"Is this precautionary," a deputy asked, "or corrective?"

"Precautionary," the analyst replied. "Aligned with updated sensitivity guidance."

No one asked where that guidance came from.

Housing and urban development felt it next, indirectly. Not through climate models, but through insurance availability. A regional pilot program failed to attract underwriters willing to lock rates beyond a decade.

"Markets are becoming more conservative," someone said.

No one disagreed.

Agricultural policy followed the same pattern. Subsidy models assumed regional reliability that no longer lined up precisely with planting data. The solution wasn't reform—it was flexibility.

Flexible timelines.

Flexible thresholds.

Flexible expectations.

Flexibility cost money, but it didn't look like loss. It looked like prudence.

An internal economic review captured it best, though no one highlighted the sentence.

Increased uncertainty introduces friction across long-horizon commitments, resulting in incremental cost layering rather than discrete shocks.

Incremental was the word that made it survivable.

Defense procurement felt it without naming it. Long-term platform assumptions required slightly higher contingency buffers. Navigation redundancy budgets expanded quietly. Nothing canceled. Nothing delayed.

Just more margin.

A senior official summarized it during a closed session.

"We're not seeing risk," he said. "We're seeing inefficiency."

Inefficiency was tolerable.

Inefficiency could be funded.

The economist who had first raised the issue watched how the language evolved.

At first, she had spoken about *pricing distortion.*

Now people said *risk premium normalization.*

Same meaning.

Safer phrasing.

She tried once to draw the line explicitly.

"These costs aren't independent," she said. "They're linked to the same underlying assumptions."

The response was immediate but gentle.

"We understand the linkage," the chair said. "But from a policy perspective, these are distributed impacts."

Distributed meant no owner.

Later, in a smaller group, someone asked the question no one wanted to document.

"What happens if this continues?"

The economist chose her words carefully.

"Then predictability becomes expensive."

Silence.

"That's not collapse," she added. "It's drag."

Drag was the right word. Engineers understood it. Economists tolerated it. Politicians rarely noticed it until elections were long past.

A budget analyst asked, "Is there a threshold where this forces revision?"

She shook her head. "Not a clean one."

"Then we manage it," he said, relieved.

Management again.

The memo was updated—not revised, just appended. A new section on *long-horizon economic sensitivity* appeared, carefully cross-referenced to existing uncertainty language.

It said nothing new.

It changed everything downstream.

From that point on, any increase in cost could be justified as alignment with long-horizon realism. Any delay could be framed as responsible caution. Any inefficiency could be explained as resilience.

No one lied.

No one panicked.

But something subtle happened inside government awareness: the future stopped being something you optimized for and became something you buffered against.

Jun noticed it during a briefing when a senior planner said, almost offhandedly, "We're planning for variability now, not outcomes."

That sentence passed without comment.

Later, Lena said quietly, "That's expensive."

Paul nodded. "And invisible."

Because budgets absorbed it.

Costs diffused across categories that drew scrutiny. No single line item ever seemed large enough to question.

Because taxpayers never saw a line item labeled *loss of predictability*.

Because no single decision caused it.

It was the cost of keeping assumptions intact just a little longer.

The unspoken economic problem wasn't recession.

It wasn't collapse.

It wasn't even instability.

It was that the price of pretending the future behaved like the past had begun to compound.

And compounding costs didn't demand explanation.

They just demanded accommodation.

Government awareness deepened another notch—not into alarm, but into acceptance.

From here on, the system wouldn't ask whether assumptions were still true.

It would ask how much it cost to keep using them.

And as long as the answer remained *affordable*, nothing else would change.

∝ ∝ ∝

The acceptance phase didn't come with a memo.

It arrived as a tone.

Budget meetings grew longer without growing sharper.
Agendas expanded to accommodate explanation.
Slides gained additional bands, additional shading, additional explanatory notes that no one read aloud but everyone relied on. Words like *buffer*, *contingency*, and *prudence* migrated from footnotes into summaries.

A senior fiscal analyst put it plainly during one closed session.

"We are paying for uncertainty we cannot eliminate."

No one disagreed.

"What matters," she continued, "is whether that payment remains proportionate."

Proportionate to what was never specified.

The economist who had first noticed the misalignment stopped trying to unify the effects. She learned quickly that unification implied cause, and cause implied responsibility.

Instead, she tracked second-order behavior.

Municipal borrowing terms stretching.
Insurance underwriting horizons shortening.

Federal programs quietly shifting from fixed commitments to adjustable frameworks.

Each move made sense on its own.

Together, they formed a pattern no dashboard displayed.

A deputy asked one afternoon, "Is this cyclical?"

The answer came back immediately. "No."

"Structural?"

A pause.

"Emergent."

Emergent was useful. Emergent meant no one had designed it. Emergent meant no one was accountable.

The phrase spread.

An internal note described *emergent fiscal drag driven by long-horizon uncertainty normalization.*

The sentence was technically accurate. It was also the furthest the truth would be allowed to travel.

Defense planners absorbed it with professionalism. They didn't argue about the science; they adjusted the math.

Redundancy budgets expanded. Replacement timelines gained flexibility. Navigation systems were layered. Predictability had become a resource rather than a given.

One planner summarized it succinctly.

"We're building optionality."

Optionality sounded proactive. Optionality justified expense without implying danger.

Climate policy took the same shape. Targets remained intact, but implementation schedules loosened. Benchmarks became ranges. Ranges became narratives.

A policy advisor said it during a coordination call, not defensively, just factually.

"Precision is no longer realistic at this scale."

No one challenged him.

Precision, after all, was what had started to erode.

The memo—still technically about science—now appeared in economic briefings as a citation of record. It wasn't read. It was referenced.

"This aligns with the synthesis," someone would say.

And the discussion would move on.

Jun noticed that no one asked scientists for new answers anymore. They asked for confirmation that nothing had crossed an agreed line.

"How confident are you that this remains within the previously characterized envelope?"

That question appeared everywhere.

Lena answered it truthfully every time.

"We remain within it."

What she didn't say—what no one asked—was whether the envelope itself still described the world they were planning for.

The cost modeling eventually caught up to the language.

A Treasury review projected modest but persistent efficiency losses across decades. Not enough to revise growth forecasts. Enough to explain why future budgets required more flexibility than past ones.

The conclusion was careful.

These effects are diffuse and don't indicate systemic instability.

Diffuse again.

The economist read the report late one night and realized something that had been forming for months.

This was no longer about warning anyone.

The system had already chosen its response.

Not correction.

Not confrontation.

Accommodation.

Accommodation was the only response that didn't force anyone to say the old baselines were wrong.

Accommodation was elegant. It avoided conflict. It preserved continuity. It allowed governments to remain internally consistent while the ground beneath their assumptions shifted slowly, politely, almost apologetically.

She wrote one last note to herself and never sent it.

We are pricing ignorance instead of resolving it.

The next day, the working alignment group renewed its charter without discussion.

Government awareness had completed its arc.

The issue was now officially understood, officially bounded, officially affordable.

No emergency powers invoked.
No public statements revised.
No assumptions formally withdrawn.

Just a future that cost slightly more to plan for every year.

And because the cost arrived as increments rather than shocks, it would never be attributed to its source.

The Moon still drifted.
The models still ran.
The plans still balanced.

What had changed was invisible to anyone not watching closely:

The margin for surprise had been quietly spent.

From here, the story does not turn on discovery.

It turns on endurance.

How long institutions can keep absorbing friction without admitting that the assumptions they were protecting no longer describe the world they inhabit.

And how long the people inside those institutions can live with the knowledge that they saw it early—and learned, very carefully, how to look away without closing their eyes.

The shift didn't happen in a room.

It happened in drafts that stopped changing.

Public-facing reports began to look remarkably similar across agencies. Not identical — that would have been suspicious — but harmonized in a way that suggested everyone was reading from the same internal weather.

Lena noticed it when a quarterly release crossed her desk for review.

The data was solid.

The charts were honest.

The language was precise.

It also stopped just short of connection.

A section on long-term variability described widening confidence bands in careful detail, then ended with a sentence she had seen before, almost word for word:

These effects are not expected to impact near-term stability.

Near-term had become the anchor. Everything else was allowed to float.

She flagged a paragraph that mentioned coupled systems interacting over extended horizons.

The response came back quickly.

We recommend omitting this level of integration in public summaries to avoid misinterpretation.

Avoid misinterpretation.

Not *avoid panic.*

Not *avoid scrutiny.*

Just misinterpretation.

Jun saw the same pattern in a press brief he was asked to sanity-check. The reporter's questions were answered accurately, narrowly, and completely.

"Is anything changing?"

"Models are evolving, as they always do."

"Should the public be concerned?"

"There is no indication of immediate risk."

Both statements were true.

What wasn't said was also true — and far more difficult to explain in a headline.

Secondary effects were still present in the full reports, buried where specialists would find them. Long-term sections expanded quietly. Appendices grew heavier.

Anyone looking carefully could see the shape of the issue.

No one looking casually would.

Paul pointed it out during a late call.

"Notice how nothing gets removed," he said. "It just moves."

Lena nodded. "Downward. Outward. Later."

"Exactly," Paul said. "Time becomes a filter."

Funding followed the same logic.

Research budgets weren't cut. That would have raised questions. Instead, they were reaffirmed — even expanded — with language emphasizing continued investigation and refinement.

A program officer said it plainly in a closed session.

"We want to stay ahead of the science."

Ahead didn't mean public.

It meant insulated.

Jun received notice that his project had been renewed for another cycle. No changes requested. No redirection.

Just quiet support.

That, more than anything, told him how the decision was being made.

If the goal had been suppression, funding would have dried up. If the goal had been exposure, messaging would have shifted.

Instead, the system chose a third option: keep the work going where it couldn't surprise anyone.

Ruth noticed it when she was asked to review a draft op-ed written by a senior official.

The piece praised scientific rigor, long-term monitoring, and the value of patience in complex systems. It didn't mention the Moon. It didn't need to.

She suggested adding a line about systemic coupling.

The suggestion was thanked.

It wasn't adopted.

No one told her not to push.

The pattern was clear now.

Public reports remained technically accurate.
Secondary effects were consistently framed as long-term.
Press summaries avoided system-level interaction.
Research continued—quietly, steadily, generously.
Messaging returned, again and again, to the same phrase.

No immediate risk.

Ruth said it out loud once, not as an accusation, just as an observation.

"There's no memo saying we don't tell the public."

Paul nodded. "There wouldn't be."

Jun added, "There doesn't have to be."

Because by this point, everyone understood the cost of saying more.

In consequences.

A single sentence, framed poorly, could force revisions to plans that had taken years to align. It could trigger questions that had no clean answers. It could turn a managed uncertainty into a public demand for certainty.

And certainty was the one thing no one could offer.

So the decision made itself.

Through triage.

What must be said now.
What can be said later.
What is better left in the hands of specialists.

Lena realized it one evening as she prepared remarks for a briefing she knew would be quoted.

Every sentence she wrote was true.

Every sentence she deleted was also true.

The difference was timing.

And timing, in a system built to preserve continuity, was everything.

By the time the practice stabilized, no one remembered choosing it.

It simply became how things were done.

No announcement.

No directive.

Just a shared understanding that some truths were better managed than shared—institutions couldn't afford to react to them yet.

The decision was never written down.

It lived in the space between accuracy and implication.

And once it settled there, it was remarkably hard to dislodge.

The Decision That Is Never Written Down

After that, nothing dramatic happened.

There was no meeting convened to decide what the public should know. No closed-door vote. No policy document circulated for comment. The systems in place simply continued to do what they had already begun to do—filter, compress, contextualize.

Reports moved up the chain unchanged in substance but altered in posture. The numbers stayed where they were. The margins of error were preserved. The language surrounding them softened, not through instruction but through repetition. Phrases that provoked follow-ups were replaced by ones that concluded discussion. Not falsehoods—never that—but endpoints.

What had once been flagged as interaction became correlation. What had once been described as coupling became coincidence. What had once been labeled emergent became deferred.

No one could point to the moment it happened. It was visible only in retrospect, as a drift so gradual it felt intentional only after it had already settled.

Drafts came back with light edits.

"Clarify scope."

"Reduce speculative framing."
"Emphasize established timelines."

None of those notes were unreasonable. Each could be defended on methodological grounds. Each made the document cleaner. More responsible. More resistant to misinterpretation.

And each removed just enough connective tissue to prevent the reader from seeing the system as a system.

The press briefings followed naturally. They always did. Communications teams learned which formulations produced stability and which produced volatility. The phrase *no immediate risk* emerged because it worked. It ended questions without appearing evasive. It reassured without promising. It implied vigilance without urgency.

Once it entered circulation, it began to appear everywhere.

No immediate risk to infrastructure.
No immediate risk to public safety.
No immediate risk requiring action at this time.

The qualifier did all the work.

Behind the scenes, funding approvals continued to arrive. Not loudly. Not ceremonially. Grants were renewed. New lines appeared in budgets under existing headings. Review panels recommended further study with unusual unanimity. Whatever disagreement existed about interpretation vanished when it came time to decide whether the work should continue.

It should. Everyone agreed on that.

The research was too important to interrupt. Too complex to rush. Too embedded in long-term planning to be subjected to reactive oversight.

So it was protected the only way institutions know how to protect something delicate: by normalizing it.

By the time anyone noticed the pattern, it no longer felt like a choice.

It felt like professionalism.

The shift became measurable only when someone bothered to compare versions.

Not line by line—that would have shown nothing. The edits were too small, too defensible. It was only when Lena stacked six months of public summaries against their internal counterparts that the shapes appeared. The facts were still there, unmistakable, clear, but they no longer touched each other. Each stood alone, properly bounded, stripped of the implication that they belonged to the same system.

She didn't mention it at first. She wasn't sure what she would be accusing anyone of. No rule had been broken. No standard violated. If anything, the documents were cleaner than before. Less speculative. More careful.

Careful had become the virtue of the moment.

In internal discussions, the language remained precise. Models were updated. Scenarios were run. Sensitivity analyses grew thicker, more granular, branching outward in quiet confidence. The long-term curves were still discussed—privately, rigorously, without euphemism. The coupling effects hadn't vanished; they had simply retreated to rooms without cameras.

Everyone understood why. No one needed it explained.

A system under scrutiny learns to speak in layers. What must be said now. What can wait. What requires context that cannot survive compression into a headline or a hearing clip. The danger wasn't that the public would misunderstand. It was that they would understand partially, and act fully.

Paul put it bluntly one afternoon, not in a meeting, just standing by the coffee machine, watching the machine struggle with a cup it didn't recognize.

"You can't ask institutions to react proportionally to uncertainty," he said. "They only know how to react to certainty. And certainty, in this case, would be catastrophic."

Lena knew what he meant. Certainty triggered mandates. Mandates triggered timelines. Timelines collapsed nuance. And once nuance was gone, everything became urgent by default.

Urgency was expensive.

So the work continued, insulated not by secrecy but by cadence. Updates released on schedule. Findings framed conservatively. The absence of alarm became, itself, a signal of control. Markets stayed calm. Legislators found other crises to occupy them. The system exhaled, believing it had bought time.

And time, at least, was something researchers could use.

Jun noticed it from a different angle, not in the data itself but in the way decisions were beginning to arrive fully formed. His inbox was filled with invitations that used to be phrased as questions. Now they were statements framed as conclusions. *We'll be proceeding with the following assumptions. For planning purposes, we're treating secondary effects as post-horizon.* No one asked him to object. They assumed he wouldn't—or that if he did, it would be quietly absorbed and categorized as a minority position.

He didn't object. The assumptions were, technically, defensible. The horizon was a choice. All horizons were. Choose one far enough out and everything dangerous became theoretical.

That was the trick. Make the risk real but not immediate. Important but not actionable. Urgent for specialists, invisible to everyone else.

By the time external reviewers began to echo the language unprompted, the practice had fully stabilized. Journalists used the same phrasing without realizing where it had originated. Policy briefs mirrored the tone because it sounded responsible. Cautious. Adult.

No one could point to the author of the shift because there wasn't one.

What followed wasn't escalation, but dispersion.

It had emerged from alignment—of incentives, of fears, of an unspoken agreement that preserving the capacity to act later required restraint now. Not silence. Restraint.

The decision was never written down.

It didn't exist as a policy. It existed as a gradient—one that every participant learned to feel, adjust to, and eventually navigate without thinking. And once that happened, reversing it would have required something far more disruptive than any of the truths it had set aside.

It would have required saying, out loud, that the system had chosen timing over transparency.

And no one was prepared to be that honest yet.

Lena finally said something when she realized the questions were changing.

They were no longer asking *what does this mean?* They were asking *how long do we have before it matters?*

The difference was subtle, but it carried weight.

She brought it up in a small room with bad acoustics and a table that had seen too many budget fights. Ruth was there, flipping through a printed deck she didn't really need. Paul leaned back in his chair, arms crossed. Jun sat forward, hands folded, already listening.

"They're not wrong," Lena said, carefully. "But they're skipping steps."

Ruth looked up. "Which steps?"

"The ones where we admit the interactions don't wait for permission," Lena said. "We're treating coupling like a background condition instead of a driver."

Paul exhaled through his nose. "Because the moment we call it a driver, someone asks what to do about it."

Jun nodded. "And we don't have an answer that fits in a sentence."

Ruth closed the folder. "We have answers. We just don't have actions."

"That's worse," Paul said. "Actions can fail. Answers without actions just sit there and make everyone nervous."

Lena hesitated. "So we just… keep framing it out?"

Ruth didn't answer immediately. When she did, her voice was steady. "We keep framing it *appropriately.*"

"For whom?" Lena asked.

"For the system we're inside," Ruth said. "Which includes us."

Jun frowned. "That's circular."

"Yes," Ruth said. "That's how systems work."

The room went quiet. Not uncomfortable—familiar. The silence of people who had already run the argument forward and backward and found the same wall every time.

Paul broke it. "Look, nobody's saying we lie. We don't lie. Ever. But we're not obligated to lead with the most destabilizing interpretation."

Lena looked at him. "Destabilizing to whom?"

"To everything," he said. "Funding. Planning. Public trust. You don't pull a thread unless you're prepared for what unravels."

"And if the thread unravels anyway?" she asked.

"Then we say we couldn't have known," Paul replied. "Which will be true, in the only way institutions understand truth."

Jun shifted in his chair. "There's a cost to that."

"There's a cost to everything," Ruth said. "This one buys time."

"For what?" Lena pressed.

Ruth met her gaze. "For understanding. For preparation. For not being forced into decisions by people who don't understand the system but will have authority over it the moment it becomes a headline."

The word *headline* hung there.

Lena leaned back, feeling the familiar resistance in her chest—the one that wasn't moral outrage so much as intellectual discomfort. "We're assuming the public can't handle uncertainty."

"No," Jun said quietly. "We're assuming institutions can't."

Paul glanced at him. "That's the first honest sentence anyone's said today."

Ruth didn't smile. "Honest sentences are expensive. They ripple."

"So we smooth them," Lena said.

"So we *sequence* them," Ruth corrected. "There's a difference."

Lena wasn't sure there was. But she also wasn't sure she wanted to test it.

Later, in the hallway, Jun caught up to her.

"You're not wrong," he said.

"I know," she replied.

"But you're early."

She stopped walking. "Early to what?"

"To the part where this becomes a choice again," he said. "Right now, it's momentum. Momentum feels inevitable until it doesn't."

"And when it doesn't?"

Jun shrugged. "Then someone will have to say the thing we're all avoiding."

No one said it.

Everyone recognized it as inevitable.

Lena watched a group of staffers pass, laughing about something unrelated, blissfully inside the version of the world that still fit together cleanly.

"Who?" she asked.

Jun looked at her for a moment too long. "Whoever runs out of patience first."

That night, Lena rewrote the briefing.

She didn't change the conclusions. She didn't remove any data. She adjusted verbs. She moved a paragraph. She took one sentence about interaction effects and relocated it to an appendix she knew would be read by exactly three people.

Every change was defensible.

Every change made the story harder to see.

The document remained accurate. Its implications grew easier to set aside.

When she finished, she stared at the screen, cursor blinking at the end of a document that would be quoted, summarized, compressed, and circulated until it lost all resemblance to the room it had been born in.

She didn't feel dishonest.

She felt complicit.

And the most unsettling part was how quickly that feeling faded, replaced by something smoother. Something that looked, from the outside, exactly like professionalism.

The decision was never written down.

But it was being rehearsed, line by line, in rooms like this—until no one remembered how else to speak.

∝ ∝ ∝

The pushback finally arrived from someone who wasn't supposed to care.

It happened during a cross-agency review that was meant to be procedural—status updates, budget alignments, nothing that would make a transcript worth rereading. The room was larger this time, with nameplates instead of first names, and a few people who had learned to ask questions for a living.

Midway through Lena's presentation, a man from oversight—new, sharp, still unaccustomed to the soft edges of institutional language—raised his hand.

"I'm trying to reconcile something," he said. His tone was polite. Curious. "You've described interacting systems, each stable in isolation. But your public summary treats the effects independently. Why?"

The room didn't tense. It *stilled.*

Lena answered carefully. "Because interaction doesn't automatically imply acceleration. At the scales we're discussing, it's appropriate to treat secondary effects as long-term."

The man nodded. "Appropriate for modeling, sure. But for communication?"

Paul leaned in before Lena could continue. "Communication isn't modeling," he said. "It's risk management."

"Whose risk?" the man asked.

Paul didn't hesitate. "Everyone's."

Ruth interjected, calm as ever. "We're not withholding information. We're prioritizing interpretability. System-level interactions are difficult to convey without context, and context doesn't survive public compression."

"So the public doesn't get the full picture," the man said.

"They get an accurate picture," Ruth replied. "Just not a destabilizing one."

That word again.

The man frowned. "Destabilizing sounds subjective."

"It's empirical," Jun said. "We've seen what happens when uncertainty is presented without resolution. Markets react. Policy hardens prematurely. Research pathways narrow."

"Those are consequences of reaction," the man said. "Not of truth."

Lena felt the pull to agree—and the equal pull not to.

Paul spoke instead. "Truth doesn't exist in a vacuum. It exists in systems. And systems respond."

The man looked around the table. "So the ethical position is… what? That timing matters more than disclosure?"

Ruth folded her hands. "The ethical position is that harm minimization requires judgment. Transparency without proportionality can be reckless."

"Or paternalistic," the man countered.

The word landed harder than *destabilizing* ever had.

Silence followed. The accusation fit too well to dismiss easily.

Lena broke it. "We're not deciding what people deserve to know," she said. "We're deciding what the system can absorb without breaking."

"That's still a decision," the man said. "And it's one you're making on behalf of others."

Jun looked at Lena, then at the table. "Every institution does that," he said. "The alternative is paralysis."

"And the alternative to that," the man replied, "is accountability."

Ruth's voice softened. "Accountability requires agency. Agency requires options. Right now, we're still building those options."

"While the public waits," he said.

"Yes," Ruth said. "Because waiting, in this case, preserves the ability to act later."

The man closed his notebook slowly. "You're assuming later is guaranteed."

No one answered.

The silence held, heavy but procedural.

No one challenged the premise.

No one accepted responsibility for disproving it.

After the meeting, the arguments continued in fragments in hallways, via emails, and on phone conversations that began with *off the record* and ended with absolutely nothing resolved. Ethics was invoked carefully, like a fragile instrument no one wanted to calibrate too precisely.

At one point, Lena found herself arguing the opposite side to a junior analyst who hadn't yet learned how expensive clarity could be.

"If we put everything out there," the analyst said, "people will demand action."

"And if we can't act?" Lena asked.

"Then we say that."

"And then?" Lena pressed.

The analyst hesitated. "Then… we figure it out."

Lena nodded slowly. "Figure it out under pressure. With timelines set by outrage instead of evidence. With solutions chosen for visibility instead of effectiveness."

"That still feels more honest," the analyst said.

"It is," Lena agreed. "It's just not always survivable."

That night, Ruth sent an email she deleted twice before sending.

We need to acknowledge that this is no longer just a communications posture. It's becoming a moral framework.

Paul replied an hour later.

All frameworks are moral once they have consequences.

Jun's response came the next morning.

The question isn't whether we're choosing. It's whether we admit we are.

No one replied to that.

Because admission would require language they had already trained themselves not to use.

And so the arguments circled without resolution—ethics against stability, disclosure against control, truth against timing—never landing long enough to force a reckoning.

The decision remained unwritten.

But it was no longer invisible.

It was present in every pause before a sentence.
Every caveat added.
Every conclusion deferred.

And with each repetition, it grew easier to justify—not as a choice, but as the only responsible way to proceed.

Which was, perhaps, the most dangerous argument of all.

∝ ∝ ∝

The reckoning, when it came, didn't arrive as a confrontation.

It arrived as fatigue.

Weeks passed. Then months. The arguments stopped evolving and began to repeat themselves, each side polishing the same sentences until they sounded less like convictions and more like scripts. What had once felt like ethical urgency softened into procedural routine.

That was when the danger deepened.

It surfaced during a late meeting that should have been canceled. Half the room was remote, faces flattened into small rectangles, eyes dulled by the glow of documents they already knew too well.

Ruth opened with a familiar line. "We need to align on messaging for the next release."

Jun muted himself, rubbed his face, then leaned forward. "We keep calling it messaging," he said. "But this isn't spin anymore. It's doctrine."

Paul frowned. "That's dramatic."

"No," Jun said. "It's precise."

Lena looked up. "What do you mean?"

Jun took a breath. "Messaging is flexible. Doctrine isn't. Doctrine defines what questions are acceptable."

The word *doctrine* made several people shift in their seats.

Ruth responded evenly. "We're defining scope, not belief."

Jun shook his head. "Scope becomes belief once it's enforced socially instead of formally."

Paul crossed his arms. "No one's being punished for saying more."

"Not overtly," Jun said. "But watch what happens when someone does."

Silence.

They all had.

A junior modeler had raised concerns two weeks earlier—nothing reckless, just an email noting that certain interaction terms, while still long-term, were beginning to converge faster than expected. The email had been forwarded. Then quietly parked. No reprimand. No correction. Just an absence of response so complete it taught its own lesson.

Lena spoke slowly. "We're training people."

"To be careful," Paul said.

"To be quiet," Jun corrected.

Ruth closed her eyes for a moment. "Careful and quiet aren't the same thing."

"They are when the cost of speaking isn't discipline," Jun said, "but isolation."

That landed.

A voice from the screen—oversight again, the same man—cleared his throat. "May I ask something directly?"

Ruth nodded.

"When this started," he said, "did any of you believe it would last this long?"

No one answered immediately.

Lena finally said, "We believed it would buy time."

"And what did you think you'd do with that time?" he asked.

Another pause.

Paul answered. "Understand the system well enough to act without guessing."

"And if that point keeps moving?" the man pressed.

Jun didn't hesitate. "Then we're not buying time. We're spending it."

Ruth leaned forward. "You're assuming we're standing still."

"I'm assuming," the man said, "that silence feels safer than escalation. And that safety is addictive."

The word *addictive* hung in the air, unwelcome but accurate.

Lena felt something tighten in her chest. "What's the alternative?" she asked. "We dump everything? Force premature action? Break trust when we can't follow through?"

The man nodded. "Those are real risks. But so is normalizing a posture where truth is always postponed."

Paul sighed. "Truth isn't postponed. It's staged."

"For how long?" the man asked.

No one had an answer they liked.

The uncertainty lingered longer than the question.
Discomfort settled into the space where resolution should have been.
The discussion moved on without resolving either.

After the call ended, the room emptied slowly. No one rushed out. No one lingered either. It felt like the aftermath of an argument that had no winner—only costs that hadn't yet been assigned.

Lena stayed behind, packing her bag with unnecessary care. Jun waited near the door.

"They're not wrong," he said quietly.

"I know," she replied.

"But neither are we."

"That's the problem," Jun said. "If one of us were clearly wrong, this would be easier."

Lena looked at him. "Do you ever think about what happens when this ends?"

Jun nodded. "All the time."

"And?"

He hesitated. "I think about transcripts. About people reading back through statements that were accurate but incomplete, and asking why no one connected them sooner."

"Will they be right?" she asked.

Jun considered it. "They'll be late."

That night, Lena opened a draft she hadn't touched in weeks.

It wasn't for public release. It wasn't even scheduled. It was a private document—notes, really—where the interactions were laid out plainly, without qualifiers, without timing language. A map of the system as it actually behaved.

She stared at it for a long time.

Then she added a date to the filename.

Not as a plan.

As a marker.

Because somewhere beneath the triage, beneath the professionalism, beneath the rehearsed restraint, she wanted proof that the full truth had existed—if only in waiting.

The decision was still never written down.

But the cost of that decision was no longer abstract.

It was accumulating quietly, like interest.

And everyone in the room, whether they admitted it or not, could feel the balance shifting—toward a future where timing would no longer be theirs to control.

And everyone in the room, whether they admitted it or not, could feel the balance shifting—toward a future where timing would no longer be theirs to control.

That awareness changed how they spoke.

Statements narrowed.

Certainty was handled more carefully.

Silence became a tool rather than a gap.

Not what they said—what *preceded* it. The pauses grew longer. The caveats more deliberate. Sentences were framed as if they might one day be replayed, lifted out of context, examined by people who wouldn't remember how uncertain everything had been.

Ruth was the first to acknowledge it openly.

"We're running out of neutral ground," she said during a smaller meeting, the kind where no one took notes because everyone assumed they'd remember. "Soon, not choosing will start to look like a choice."

Paul leaned forward. "It already does—to people who want it to."

"That's not who I'm worried about," Ruth replied. "I'm worried about the people who *don't* want it to. The ones who still believe this is just analysis."

Jun tapped a pen against the table. "Analysis ends when it starts shaping behavior."

Lena nodded. "And it already is. Just not visibly."

Paul frowned. "So what are you proposing?"

Ruth didn't answer right away. When she did, her voice was careful. "I'm saying we need to be honest about what we're doing—even if we don't announce it."

"That's a contradiction," Paul said.

"No," she said. "It's an internal one."

Jun raised an eyebrow. "You want to name the thing without naming it."

"I want us to stop pretending this is still accidental," Ruth said. "Because if we keep calling it drift, we'll lose the ability to steer it."

Lena felt a chill. "Steer it toward what?"

"Toward survivability," Ruth said. "For the work. For the institutions. For the public, eventually."

Paul shook his head. "That's still a strategy."

"Yes," Ruth said. "And pretending it isn't one doesn't make it less real."

The word *strategy* sat uneasily between them. It implied intent. Design. Ownership.

Jun broke the silence. "Once we admit that, the ethical calculus changes."

"How?" Paul asked.

"Because then we're no longer managing uncertainty," Jun said. "We're managing access to understanding."

No one liked that phrasing.

Lena spoke softly. "We already are."

Paul looked at her. "Are you saying we've crossed a line?"

She considered it. "I'm saying the line moved. Quietly. And we followed it because staying put felt irresponsible."

Ruth exhaled. "That's the most dangerous part."

"What?" Paul asked.

"When responsibility starts to mean containment," Ruth said. "Instead of disclosure."

The room went still again. Not with shock. With recognition.

Jun leaned back. "The irony," he said, "is that everything we're doing is defensible. Every step can be justified on its own merits."

"That's how it works," Lena said. "No single act carries the weight of the whole."

Paul looked from face to face. "So what's the alternative? We blow it open? Force synthesis into places that aren't equipped to handle it?"

"No," Ruth said. "Not yet."

"Then what?" Paul pressed.

Ruth met his gaze. "Then we accept that the decision *has* been made—not about truth, but about *when* truth is allowed to become inconvenient."

No one argued.

Because they all understood that inconvenience, at scale, was indistinguishable from crisis.

Lena thought again of the private draft on her laptop. Of the date in the filename. Of the quiet hope that someday, someone would read it and understand that the full picture had not been lost—only delayed.

Delay felt different now.

It no longer felt like patience.

It felt like preparation for something they could sense approaching but couldn't yet name.

And in that unnamed space—between urgency and restraint, between accuracy and consequence—the decision that was never written down took on its final shape.

Not as silence.

But as placement.

And once that logic took hold, there was no returning to a world where information simply flowed wherever truth led it.

The balance had shifted.

And everyone in the room knew that whatever came next would no longer be about *whether* to speak—but *where* the truth would be allowed to live.

∝ ∝ ∝

The realization settled slowly.

There was no meeting where it was named, no document that recorded the moment. No email that could later be pointed to and circled in red. The shift announced itself only in retrospect, visible the way erosion is visible—by comparison, by absence, by the way something familiar no longer quite matched its earlier outline.

What they had built wasn't silence, and not deception. It wasn't a lie, and it wasn't even secrecy in the traditional sense. Information still moved.

Reports were still written. Data still existed, intact and internally consistent. What had changed was the posture toward it—the angle at which it was held.

Information became something that could be delayed without being denied, shaped without being falsified. It could be framed as incomplete, premature, insufficiently contextualized. It could be acknowledged and then set aside. It was inconvenient to act on too quickly. Each individual decision remained defensible on its own terms. Each delay could be justified. Each qualification could be explained. Each hedge sounded responsible.

Together, they formed something heavier than intent.

No one could point to the moment responsibility became management rather than action or choice. There was no line where stewardship crossed into control. No single sentence where caution hardened into policy. The vocabulary never changed enough to call attention to itself. Words like *monitor*, *review*, *reassess*, *coordinate* continued to circulate, doing the quiet work of slowing momentum without ever appearing to stop it.

No one could say when caution became containment, or who first noticed the difference.

Only that the shift had occurred—and that reversing it would now require a disruption larger than any single truth they had deferred. By the time the pattern was visible, it was already reinforced by process. Calendars were full. Briefings were scheduled. Language had been standardized. There were templates now, and expectations, and a shared understanding of how certain findings should be discussed and where they should be routed.

The decision had never been written down because it didn't exist as an order.

It existed as alignment.

As habit.

As the quiet consensus that understanding must sometimes wait for permission.

People learned, quickly and without instruction, what kinds of questions generated traction and which ones dissolved into polite acknowledgment. They learned how to phrase concerns so they would be logged rather than escalated. They learned when to pause, when to soften, when to substitute uncertainty for insistence. None of this was taught. It emerged naturally, the way organizational gravity always does, pulling behavior into stable orbits.

By the time anyone thought to question it openly, the system was already responding—not to facts, but to their implications. Raw measurements mattered less than what they suggested might follow. Projections became the point of tension because they were difficult to accommodate. The farther forward a conclusion reached, the more carefully it was handled.

This wasn't suppression. Suppression implied force. This was filtration.

And so the next step wasn't announced, debated, or authorized.

It simply followed.

If truth couldn't yet move freely, it would have to be placed—carefully, professionally, in spaces designed to hold it without letting it spread too quickly or too far. Restricted briefings. Closed sessions. Technical appendices few people read. Repositories that required justification to access because its timing was considered sensitive.

That wasn't secrecy.

It was containment.

And like all containment, it carried with it the assumption that someone, somewhere, would know when it was safe to release what had been held. That judgment, too, was never assigned. It passed silently from role to role, embedded in procedures, deferred along chains of responsibility until it no longer belonged to anyone in particular.

The system continued to function smoothly. Meetings ended on time. Reports were filed. No alarms were triggered.

Budgets cleared review. Calendars remained full. Interim conclusions were circulated with careful wording and familiar caveats, then acknowledged and set aside. Each step appeared ordinary in isolation, indistinguishable from the thousand small acts of maintenance that kept institutions running.

The data remained intact, preserved with the same rigor as always—measured, archived, verified, and protected against loss. Nothing was altered. Nothing was erased.

Only its movement had changed.

Where the data once flowed outward by default, it now paused. Where it had circulated freely, it now passed through hands trained to slow it, frame it, and decide when it was appropriate to proceed. Not stopped. Not hidden. Simply… held.

Nothing had gone wrong.

That was the problem.

The system didn't recognize this as a failure. From its perspective, everything was working.

∝ ∝ ∝

THE CONTAINMENT STRATEGY (HOW INFORMATION IS SILOED)

Information didn't disappear.

It was *sorted.*

That distinction mattered to everyone involved. Nothing was classified. Nothing was redacted. Everything remained accessible—just not *collectively*, and not *casually*. The system didn't choke the flow of knowledge; it routed it.

Academic journals became the primary reservoir. The work went there first, passing through review processes that appreciated restraint and punished overreach. Papers grew narrower. Conclusions tightened. Uncertainty sections lengthened, absorbing what couldn't yet be reconciled.

The signal was preserved, but fragmented—distributed across subfields that rarely spoke in the same room.

Paywalls did the rest, quietly. Not as barriers, exactly, but as filters. Access required intent. Time. Institutional affiliation. The people who cleared those thresholds were the ones already trained not to extrapolate too far, not to generalize without explicit permissions or prior approvals. They read carefully. They argued responsibly. They published follow-ups that stayed within the same lanes.

Conferences became echo chambers of fluency. Presentations assumed shared context and rewarded modesty. Questions sharpened methods but dulled implications. When someone gestured toward synthesis, it was acknowledged politely and deferred to "future work." Future work was always welcome. Future work never demanded action.

Appendices swelled. Supplemental materials multiplied. Summary pages thinned as explanation accumulated elsewhere.

What remained above was just an overview, not substance. Entire dynamics lived there—mathematically explicit, carefully caveated, technically impeccable. They were cited, discussed, admired. Rarely quoted.

∝ ∝ ∝

None of this felt restrictive. On the contrary, it felt like freedom. Scientists spoke openly, confidently, knowing the boundaries were professional rather than imposed. No one told them not to speculate; they simply learned which forms of speculation survived peer review and which didn't.

Language adapted.

Extrapolation became "outside scope."

Synthesis became "premature."

System-level statements were reframed as "hypotheses requiring further validation."

The norms enforced themselves.

∝ ∝ ∝

Lena noticed it when she reviewed a paper she might once have questioned more aggressively. The methods were sound. The conclusions were careful. The implications—left unstated—were enormous. She approved it without comment.

Jun saw it when a colleague presented results that clearly interacted with another group's work and chose not to mention the overlap. Not because it wasn't real, but because it wasn't *theirs*.

Ruth recognized it when media requests began to change tone. Reporters still asked questions. They were still answered. But the answers stayed within domains, within expertise, within lines that editors understood how to trim without distorting.

The mass audience never encountered the whole picture.

No single outlet, discipline, or voice was permitted to assemble it.

And the system rewarded this restraint. Careers advanced. Grants renewed. Reputations solidified around rigor and caution, rather than urgency, risk, or alarm. Those who attempted synthesis too publicly found themselves labeled speculative, alarmist, or—worse—imprecise.

Precision became the virtue that kept truth in its place.

Paul described it once, not unkindly, during a hallway conversation. "We didn't silence anyone," he said. "We just made sure the full sentence could only be spoken in fragments."

∝ ∝ ∝

Lena didn't respond. She was thinking about how effective it was. How elegant. How completely aligned with the values they claimed to hold.

The containment strategy had no author. It required none.

It worked because it relied on existing incentives, professional pride, and the deeply ingrained belief that knowledge must be handled carefully lest it be misunderstood. Every safeguard was reasonable. Every boundary defensible.

The result was a landscape where information lived—rich, detailed, and intact—but nowhere that it could gather enough momentum to force a reckoning.

The system worked exactly as designed.

∝ ∝ ∝

And as long as it did, the most consequential truths would remain visible only to those trained not to raise their voices.

That wasn't suppression.

It was stewardship.

And everyone involved knew, even if they never said it out loud, that once information learned how to stay contained, it rarely remembered how to escape.

"It's elegant," Paul said, tapping the side of his tablet. "I'll give you that."

Lena looked up. "Elegant isn't a defense."

"It's not an accusation either," he replied. "It's an observation. The system's doing what systems do best—routing complexity away from pressure points."

Jun leaned back in his chair. "Pressure points are where accountability lives."

Paul snorted. "Pressure points are where nuance goes to die."

Ruth, seated at the head of the table, raised a hand—not to stop them, just to slow the pace. "Let's be clear about what we're arguing," she said. "No one is claiming information is being hidden."

"Functionally, it is," Jun said.

"No," Ruth corrected. "It's being *localized*."

Lena frowned. "That's a euphemism."

Ruth met her gaze. "So is 'containment.' Pick your poison."

Jun folded his arms. "Here's my problem. If the only people who can see the full system are the ones professionally disincentivized from speaking about it holistically, then we've engineered silence without ever saying the word."

Paul shook his head. "You're assuming synthesis is a moral obligation."

"Isn't it?" Jun shot back.

Paul paused. "It's an intellectual one. Those aren't the same thing."

Lena interjected, voice sharper than she intended. "They become the same thing when consequences extend beyond the lab."

Ruth leaned forward. "And they become irresponsible when synthesis outruns certainty."

"So we wait for certainty?" Lena asked. "In a coupled system?"

Ruth didn't answer immediately.

"That's not a real option," Jun said quietly.

"No," Ruth agreed. "But neither is unleashing partial understanding into systems that will act as if it's complete."

Paul nodded. "That's the core tension. You give policymakers half a picture, they'll treat it like the whole thing. And then we're locked into responses we can't walk back."

"And if we give them the full picture?" Lena asked.

Paul looked at her. "They'll still oversimplify it. But with more confidence."

The room went quiet. The arguments were no longer new.

A junior researcher at the far end of the table finally spoke. "Can I ask something?"

Ruth gestured for him to continue.

"If this is about preventing misinterpretation," he said carefully, "why aren't we focusing on *better* interpretation instead of less access?"

Paul answered before anyone else could. "Because interpretation scales badly."

"That's… cynical," the researcher said.

"It's empirical," Paul replied.

Jun turned to the researcher. "You're assuming the public discourse environment rewards careful reasoning. It doesn't. It rewards clarity, certainty, and conflict."

"So we protect the truth by isolating it?" the researcher asked.

Lena winced. "That's not how I'd phrase it."

"But that's how it feels," he said. "Like we're curating who gets to understand."

Ruth spoke gently. "We're curating *when* understanding becomes actionable."

The researcher hesitated. "Who decides that?"

No one answered.

After the meeting, Lena and Jun walked together toward the elevators.

"You notice how no one talks about intent anymore?" Jun said.

Lena nodded. "Intent is messy. Outcomes are cleaner."

"That should worry us."

"It does," she said. "But not enough to stop."

Jun stopped walking. "That's the line, isn't it?"

Lena looked at him. "Which one?"

"The one where knowing something is wrong isn't sufficient to change it."

She considered that. "Maybe that's always been true."

Jun watched the elevator doors slide shut without them. "Or maybe we've just gotten better at building systems that make it feel inevitable."

Later that evening, Paul dictated a response to a media inquiry.

He answered every question asked.

He cited peer-reviewed sources.

He stayed firmly within his domain.

When he finished, he read it back once, nodded, and sent it.

Nothing he'd said was misleading.

Nothing he'd said would allow a reader to assemble the whole.

And as the message left his outbox, he felt the familiar, unsettling mix of relief and unease—the sense that the system was holding, exactly as designed.

Which was the problem.

Because systems that hold too well rarely notice when the pressure inside them starts to climb.

∝ ∝ ∝

The pressure announced itself in small ways.

Not leaks. Not exposés. Questions.

They arrived as emails first—carefully worded, politely sourced, sent by people who clearly knew how to read between appendices. A journalist with a background in science policy and ethics. A congressional staffer with a PhD they didn't advertise. An editor asking why three unrelated papers were suddenly citing the same uncertainty language verbatim.

Each question was answerable.

Each answer was true.

Each response stayed inside its lane.

But the volume increased.

Lena noticed it when she opened her inbox one morning and realized she could predict which messages Paul would forward to communications before reading them.

"Same pattern," she said later, leaning against the doorframe of his office. "Different senders."

Paul didn't look up. "Patterns happen."

"They're correlating," she said.

He sighed. "You're projecting."

She crossed her arms. "You say that like it's an insult."

Paul finally met her eyes. "I say it like it's a warning."

Down the hall, Ruth was having a different conversation.

"This isn't panic," she said into the phone. "It's curiosity. There's a difference."

A pause. Then: "No, I'm not suggesting we restrict access. I'm suggesting we contextualize aggressively."

Another pause. Her jaw tightened. "Because context is the only thing keeping this from being misused."

When she hung up, Jun was standing there.

"They're getting closer," he said.

Ruth nodded. "They always do."

"So what happens when someone publishes the synthesis?" Jun asked. "Not a blog post. A real one. Peer-reviewed. Clean."

Ruth didn't answer immediately. "Then we treat it like any other paper."

Jun raised an eyebrow. "Which means?"

"We critique it. We replicate it. We argue about assumptions."

"And if it holds?"

"Then it becomes part of the literature."

Jun shook his head. "You're avoiding the second-order effect."

Ruth looked tired. "I'm containing it."

"That's the second-order effect," Jun said.

The first crack came from an unexpected place.

A panel discussion—livestreamed, but lightly attended—where a mid-career researcher decided to answer a question too fully. Not irresponsibly. Just… honestly.

"Well," she said, after a brief hesitation, "if you consider the interactions together, the timeline compresses in ways we haven't been emphasizing publicly."

The clip circulated.

Not virally.

Persistently.

By the time Lena saw it, it had already been framed six different ways—some cautious, some alarmist, some wildly off-base. None of them were entirely wrong.

Paul watched it twice, then closed his laptop. "We're going to get calls."

"We already are," Lena said.

Jun joined them, phone still in hand. "She didn't say anything incorrect."

"No," Ruth said from the doorway. "She said something *complete*."

They stood there for a moment, the four of them, listening to the hum of the building settling around them.

"So this is it?" Paul asked. "The breach?"

Ruth shook her head. "No. This is pressure testing."

"And if it keeps happening?" Lena asked.

Ruth met her gaze. "Then the architecture either adapts… or fails."

Jun looked at the paused frame on the screen—the researcher mid-sentence, eyes steady, voice calm. "We trained everyone not to speak that way," he said. "And still, eventually, someone does."

Lena felt a strange surge of relief. "Because truth wants to move."

Paul grimaced. "Because people get tired."

Ruth said nothing. She was thinking about how many more fragments it would take before someone assembled them into something that demanded response.

That night, Lena added another note to her private file.

Containment relies on fatigue as much as discipline.

She saved it, closed the document, and wondered—not for the first time—whether the system was holding because it was strong, or because no one had yet leaned hard enough in the right place.

Either way, the conversations were changing.

And once conversations changed, containment was no longer just an architecture.

It was a race.

∝ ∝ ∝

The race wasn't toward disclosure.

It was toward *framing*.

"Everyone's circling the same question," Paul said, pacing now, unable to sit still. "They just haven't decided how to ask it yet."

Lena watched him move. "They're asking it. Just not to us."

"That's worse," he replied. "When the question solidifies somewhere else, we lose control of the answer."

Jun looked up from his phone. "We never had control. We had *latency*."

Ruth closed the door to the conference room, shutting out the hallway noise. "Enough. We need to decide how we respond if synthesis breaks containment."

Paul stopped pacing. "Respond how?"

"Publicly," Ruth said. "Preemptively."

Lena stiffened. "That's escalation."

"It's stabilization," Ruth countered. "If someone else frames it first, we're reacting. That's the one position we've been trying to avoid."

Jun frowned. "So we synthesize it ourselves?"

Ruth shook her head. "No. We acknowledge that synthesis exists without endorsing a single narrative."

Paul let out a short laugh. "That's political language."

"Yes," Ruth said. "Because this is no longer just scientific."

Lena felt the weight of that settle. "If we do that, we're admitting the containment strategy."

"We're admitting *awareness*," Ruth replied. "There's a difference."

Jun leaned forward. "Is there? Or are we just choosing which part of the truth is survivable?"

Ruth didn't flinch. "Survivability is not a moral failure."

Jun's voice sharpened. "It can be."

Paul stepped between them verbally. "Let's not pretend the alternative is purity. The alternative is chaos shaped by whoever speaks loudest first."

A beat.

"That used to be an argument," Lena said quietly. "Now it sounds like a threat."

No one contradicted her.

A notification chimed on Jun's phone. He glanced down, then looked up slowly. "Someone just submitted a preprint."

Ruth's eyes narrowed. "On what?"

"Cross-domain interaction. Full synthesis. Clean math. No speculation language. Just… connections."

Paul swore under his breath. "Who?"

Jun named a researcher none of them could dismiss as reckless.

Lena felt her pulse quicken. "Is it public?"

"Yes."

"Peer-reviewed?"

"Not yet."

Ruth exhaled. "Then we still have time."

"For what?" Jun asked.

"To decide how the system reacts," she said. "Because it will react. The only question is whether it does so blindly."

Paul looked at Lena. "This is the moment you were waiting for."

She shook her head. "I wasn't waiting. I was hoping we'd never get here."

Jun stood. "Hope isn't a strategy."

Ruth nodded. "Neither is silence anymore."

They sat with that, the hum of the building louder now, like something under strain.

Finally, Paul spoke. "If this goes mainstream, everything changes."

"Yes," Lena said. "But maybe that's already true."

Jun looked at her. "Do you think containment failed?"

She considered it. "I think it worked exactly as designed. Long enough to make us forget what comes after."

Ruth gathered her papers. "Then we remember. And we act deliberately."

Paul hesitated. "And if deliberate action looks like manipulation?"

Ruth met his gaze. "Then we own it. Or we watch someone else do it badly."

No one liked that choice.

But it was a choice.

And for the first time since the decision that was never written down, they were being forced to acknowledge it out loud—not as drift, not as triage, not as professionalism, but as intent.

The race had entered its final phase.

Not between truth and secrecy.

But between *who* would assemble the truth first—and *whether* the system would recognize itself in the mirror when it did.

∝ ∝ ∝

They reconvened an hour later, this time without pretense.

No slides. No agenda. Just a door closed and a shared understanding that whatever happened next wouldn't stay theoretical.

Ruth started it. "We need to agree on what we're protecting."

Paul didn't hesitate. "The work."

Jun shook his head. "The system."

Lena said nothing at first. Then: "The ability to choose."

They all looked at her.

"If this becomes a binary—panic or denial—we lose agency," she continued. "Containment bought us space to think. But if we use that space only to preserve ourselves, then we deserve what comes next."

Paul rubbed his temples. "You're talking about letting go of control."

"I'm talking about redefining it," Lena said. "Control over *how* something enters the world, not whether."

Jun nodded slowly. "That's the distinction we've been avoiding."

Ruth leaned back. "And how do you propose we do that without lighting the fuse ourselves?"

Lena took a breath. "We stop pretending synthesis is taboo. We acknowledge it exists, explain why it's hard, and say—explicitly—that uncertainty is real but bounded."

Paul frowned. "That's still framing."

"Yes," Lena said. "But it's honest framing."

Jun added, "And it forces the conversation into daylight instead of letting it metastasize in fragments."

Ruth considered this. "If we do that, we're admitting that the public summaries were incomplete."

"No," Paul said. "We're admitting they were scoped."

"And people will hear 'withheld,'" Ruth countered.

Lena met her gaze. "Some will. But some already do. The difference is whether we give them a structure to argue with, or let them invent one."

Silence again. Thicker now.

Jun broke it. "There's another risk you're not naming."

Paul glanced at him. "Which is?"

"That once synthesis is acknowledged," Jun said, "we become responsible for it."

Ruth closed her eyes briefly. "Meaning?"

"Meaning we can't retreat behind domain boundaries anymore. We don't get to say 'that's not my field' when the consequences span fields."

Paul let out a breath. "That's… a lot."

"Yes," Jun said. "That's what truth costs when it finally consolidates."

Ruth opened her eyes. "And if we're wrong?"

Lena answered softly. "Then we correct. Publicly. Transparently. Like scientists are supposed to."

Paul looked unconvinced. "You're assuming the correction cycle survives first contact."

"No," Lena said. "I'm assuming it doesn't. But I'm also assuming containment won't survive it either."

That landed.

Ruth stood and walked to the window, looking out over the city—lights on, traffic moving, a world proceeding as if the future were still comfortably distant.

"We built a system that treats information like a controlled substance," she said. "Dispensed carefully. Measured doses. No sudden exposure."

Jun joined her. "And now tolerance is building."

Paul remained seated. "So what's the call?"

Ruth turned back to them. "We don't stop the preprint. We don't attack it. We don't endorse it wholesale."

Lena waited.

"We contextualize it," Ruth continued. "We publish an accompanying piece—measured, sober—that explains why synthesis is difficult, what remains uncertain, and what is *not* implied."

Paul raised an eyebrow. "And that buys us… what?"

"Legitimacy," Ruth said. "If containment is going to loosen, it does so with us still recognizable as stewards, not adversaries."

Jun nodded. "That's a gamble."

"Yes," Ruth agreed. "But every alternative is worse."

They sat with that.

Finally, Lena spoke. "Then this is the moment."

Paul looked at her. "The one we've been deferring?"

"Yes," she said. "The moment where we stop managing implications and start sharing responsibility."

No one argued.

The balance had already shifted, and arguing about it now felt like denial.

Ruth picked up her phone. "I'll make the calls."

Jun grabbed his laptop. "I'll draft language."

Paul hesitated, then stood. "I'll talk to communications. God help us."

The room held the weight of the decision without naming it. No one reached for precedent.

Silence functioned as consent.

Lena stayed seated for a moment longer, watching them move—each carrying a piece of the decision that had never been written down, now finally breaking into the open.

Containment had not failed.

It had reached its limit.

And whatever came next wouldn't be decided by architecture alone.

∝ ∝ ∝

The first draft went nowhere.

Jun sent it at 2:14 a.m.

Ruth replied at 2:27.

Paul replied at 2:41.

By morning, the document looked like it had been through a quiet argument—tracked changes layered over tracked changes, sentences tightened until they barely breathed.

Lena read it without commenting at first.

"This still sounds like we're apologizing for something we won't name," she said finally.

Paul sighed. "Because the moment we name it, we own it."

Jun didn't look up from his screen. "We already do. We're just pretending we don't."

Ruth rubbed her eyes. "Pretending buys us room."

"For who?" Lena asked.

"For everyone who isn't ready to hear the whole thing yet," Ruth said.

"And how do we know when they are?" Lena pressed.

Ruth didn't answer.

Jun broke the silence. "We keep writing as if readiness is something we can measure. It isn't. It's something we infer after the fact."

Paul looked up. "So what are you suggesting?"

Jun turned his laptop so they could see the paragraph he'd highlighted.

While recent work explores interactions across domains, such synthesis remains an evolving area of study. Interpretations should be made cautiously, with awareness of uncertainty and limits.

"That sentence," Jun said. "It does all the work. And none of the honesty."

Paul frowned. "It's accurate."

"Yes," Jun said. "And empty."

Lena nodded. "It tells people not to connect dots without admitting that the dots are already connected."

Ruth straightened. "Then rewrite it."

Jun did. Slowly.

Recent work has begun to examine interactions across domains that were previously treated independently. These interactions matter—but understanding how, when, and at what scale they matter remains incomplete.

Paul read it twice. "That's going to trigger questions."

"Yes," Jun said. "That's the point."

Ruth hesitated. "It also implies we knew this earlier."

"We did," Lena said. "Just not well enough to act."

Paul looked at her. "And now?"

"And now we know enough to say that pretending they're independent is no longer defensible," she replied.

The word *defensible* hung there, sharp.

Ruth closed her laptop. "This is the line," she said. "Once we cross it, there's no going back to containment as posture."

Jun nodded. "Only as history."

Paul leaned back, staring at the ceiling. "I hate that this feels like integrity *and* recklessness at the same time."

"That's how real decisions feel," Lena said.

A notification chimed.

Jun checked it. "The preprint's been picked up by a second outlet. Still niche, but…"

"But moving," Paul finished.

Ruth stood. "Then we move too."

They worked through the morning—rewriting, arguing, deleting phrases that felt too safe, adding qualifiers that felt just sufficient. Every sentence was a negotiation between clarity and survivability.

At one point, Paul stopped them.

"Say this goes out," he said. "Say people start connecting things we didn't explicitly connect. Say pressure builds."

"And?" Lena asked.

"And someone asks why this wasn't said sooner."

Ruth answered before anyone else could. "Then we tell the truth."

Paul blinked. "Which version?"

"The only one that matters," she said. "That we didn't know enough to act responsibly—and that when we did, we spoke."

Jun added quietly, "And that we were afraid of being wrong."

Silence followed.

Not uncomfortable this time. Grounded.

By the afternoon, the document had a title.

Not bold. Not dramatic. Just precise enough to stand.

They sent it for clearance.

No one cheered.

Lena watched the confirmation email arrive, felt the odd calm that came with having crossed a threshold you couldn't uncross.

Containment hadn't shattered.

It had thinned.

And as the words left their hands and entered a world that would read them, misread them, weaponize them, or ignore them entirely, she understood something she hadn't before.

The most dangerous part of the strategy had never been keeping information in.

It had been convincing themselves they could decide forever when to let it out.

That illusion was gone now.

What came next wouldn't be orderly.

But it would be honest.

And honesty, at this stage, was no longer a virtue.

It was a risk they had finally chosen to take.

∝ ∝ ∝

WHY THE MEDIA DOESN'T PICK IT UP

The story didn't fail because it was wrong.

It failed because it had nothing a newsroom could use.

Lena realized that when the first mainstream inquiry arrived and died in the same afternoon. The email had been polite, brief, unmistakably rushed.

We're seeing some chatter about recent preprints. Can you confirm whether this represents a change in understanding?

She and Paul drafted a response together. It took ten minutes.

"No change," Paul said, reading it back. "That's the key."

"It's true," Lena replied.

"And fatal," he said, hitting send.

In the newsroom, the editor skimmed the reply while a producer waited with a coffee already going cold.

"So?" the producer asked.

The editor shrugged. "Nothing there."

"No shift?"

"No disagreement?"

"No warning?"

The editor shook their head. "They're saying the Moon is doing what it's always done. Just… more precisely."

The producer made a face. "Do we have pictures?"

"No."

"A timeline?"

"Not really."

"Anyone yelling?"

The editor scrolled once more, just in case. "No villain. No failure. No 'experts warn.'"

They closed the tab. "It's a science explainer at best."

"What about stakes?"

The editor thought for a moment. "Abstract. Long-term. Still being studied."

The producer glanced at the rundown. "We've got a launch delay, a market wobble, and a wildfire briefing in fifteen. This isn't going to land."

"Exactly," the editor said. "It doesn't fail because it's controversial. It fails because it's boring."

∝ ∝ ∝

Elsewhere, another journalist tried harder.

She read the preprint. Then the response paper. Then two of the appendices. She understood enough to feel uneasy—but not enough to write an opening that wouldn't collapse under fact-checking.

Her notes filled with phrases she couldn't use.

System-level interaction

Emergent coupling

No immediate risk but…

Every sentence ended in ellipses.

She called Lena directly.

"I'm trying to explain why this matters," the journalist said. "And I keep running into a wall."

"What kind of wall?" Lena asked.

"One where everything important starts with 'eventually,'" the journalist replied. "Editors don't run 'eventually.'"

"That's fair," Lena said.

"So is this a story or not?"

Lena paused. "It's a record."

The journalist laughed, not unkindly. "Records don't trend."

They hung up politely.

∝ ∝ ∝

The piece never ran.

In editorial meetings, the same objections surfaced again and again, sometimes explicitly, sometimes as instinct.

There was no moment to point to.

No before and after.

No image that could anchor meaning.

No villain to simplify complexity.

No consensus phrase that could be safely repeated across platforms.

And worst of all—no action the reader could take.

Editors heard the same sentence, no matter how it was phrased:

The Moon is doing what it's always done.

More accurately measured.

More carefully modeled.

More thoroughly understood.

But still—nothing *happening.*

And in a media ecosystem trained to recognize events, not trajectories, that was the end of it.

∝ ∝ ∝

The story didn't get killed.

It simply never cleared the filters.

It slid past the desks, through the algorithms, beneath the thresholds of urgency—filed mentally alongside climate baselines, tectonic drift, background radiation.

Important.

True.

Not now.

Lena watched the coverage fail to materialize with a mixture of relief and unease.

Containment hadn't needed enforcement here either.

It had relied on something simpler.

Attention is finite.

News is episodic.

And slow truths, no matter how consequential, are almost invisible until they break something.

By the time the media would be ready to tell this story, it would no longer be about understanding.

It would be about aftermath.

And that, she knew, was the part everyone was quietly hoping to delay.

∝ ∝ ∝

The questions didn't stop.

They just changed shape.

They arrived as voicemails left after deadlines and phrased like favors. As background checks disguised as curiosity. Reporters who knew better than to pitch a headline came looking for *texture* instead—something they could test against what they already suspected.

Lena learned to recognize the sound of it.

"You're not saying anything's wrong," one reporter said over the phone. "But you're also not saying nothing's happening."

"That's because those aren't opposites," Lena replied.

A pause. The faint clatter of a newsroom behind the line.

"Off the record," the reporter said, lowering her voice out of habit, even though it meant nothing anymore. "Is this one of those stories where everyone's waiting for someone else to make it real?"

Lena chose her words carefully. "It's one of those stories where reality is ahead of language."

"That's not helpful," the reporter said, not unkindly.

"I know."

Another call, later that week.

"So let me try this a different way," a columnist said. "If I write that nothing is changing, I'm accurate. If I write that something important *will* change, I'm speculative. And if I write that institutions are quietly adjusting, I'm accusing someone of hiding something."

"Yes," Lena said.

"So which one gets me sued?"

"All of them," Lena said. "Eventually."

The line went quiet.

"That bad, huh?"

"That complex."

The reporter sighed. "Editors hate complex."

They did more than hate it. They filtered it out.

At one paper, a junior investigative reporter tried to push the story through anyway. She built a timeline—not of events, but of publications. She mapped language reuse across papers, traced funding renewals, noted the quiet consistency of phrases like *no immediate risk* and *long-term effects under study*.

Her editor skimmed it twice.

"This is interesting," he said.

"Is it publishable?" she asked.

He hesitated. "What's the nut Graf?"

"There isn't one," she admitted. "That's kind of the point."

He leaned back. "Then there isn't a story."

"What about the pattern?" she pressed.

He shook his head. "Patterns aren't news unless they break."

She tried another angle. "What if I frame it as institutional inertia?"

"Who's at fault?" he asked.

"No one," she said.

He closed the file. "Then no one clicks."

Elsewhere, rumors started doing what reporting wouldn't.

Anonymous threads appeared. Vague claims. Screenshots stripped of context. Someone posted a sentence from an appendix with the caption *they know and they're not telling you.* It got traction for a day, then dissolved into competing interpretations and louder, simpler outrage elsewhere.

Paul forwarded one of them to the group.

"This is what happens when synthesis leaks without structure," he said.

Jun replied almost immediately. *This is what happens when structure delays synthesis.*

Ruth didn't respond.

She was in a different conversation.

A senior editor at a major outlet, someone she'd known for years, didn't bother with pretense.

"Help me out," he said. "Am I missing something big?"

Ruth considered the question. "You're missing something slow."

He laughed. "That's not a thing."

"It is," she said. "It just doesn't have a clock you can put on a chyron."

"So what do I tell my readers?"

She chose honesty. "That the most important changes rarely announce themselves."

He was quiet for a moment. "You know I can't run that."

"I know," Ruth said.

"So we wait," he said.

"Yes," she replied. "Or you don't."

"And if I don't?"

"Then you'll be early," Ruth said. "And early looks a lot like wrong."

He exhaled. "You're asking me to trust institutions."

"I'm asking you to trust uncertainty," she said. "There's a difference."

He didn't sound convinced. "You realize how that sounds."

"Yes," Ruth said. "I also realize how the alternative sounds."

They hung up without resolution.

By the end of the month, there were fragments everywhere.

A podcast host hinting at "something academics won't quite say."
A newsletter linking three papers and drawing the wrong conclusion confidently.
A think-tank memo flagged as "speculative" and quietly ignored.

Noise without signal. Signal without amplification.

Lena watched it all with a growing sense of unease.

The media hadn't failed to pick up the story because it was suppressed.

It had failed because the story refused to behave.

No event.
No image.
No villain.
No clean phrasing.
No action item.

Editors weren't wrong when they dismissed it.

They were applying the filters they had.

And those filters were built for moments, not gradients.

One evening, Lena received a message from the journalist who'd called weeks earlier.

Still thinking about that wall, it read. *Feels like everyone's waiting for something to break so they can say they saw it coming.*

Lena typed back.

By the time it breaks, it won't matter who saw it first.

A few minutes passed.

That's not comforting, the journalist replied.

It wasn't meant to be, Lena typed.

She closed her laptop and sat in the quiet, thinking about how close the truth was—published, cited, available—and how completely it failed to exist in the places that shaped shared reality.

The media hadn't ignored the story.

It had done what it always did.

It had waited for the moment when subtlety would no longer be an option.

And that waiting, Lena knew, wasn't neutral.

It was a countdown no one had agreed to start.

The countdown became visible in the margins.

A producer from a Sunday show called Paul late, voice pitched low and fast. "We're not chasing a headline," she said. "We're chasing context. Off camera."

"There isn't an 'off camera' anymore," Paul replied.

"Fine. Off record, then."

Paul smiled despite himself. "That's even less real."

She pressed anyway. "Tell me this much—if I put my best team on it, am I going to end up with a story that ends in 'experts disagree'?"

"Yes," Paul said. "And you'll be right."

"Then why does it feel like I'm late?"

"Because you're not," he said. "You're early to the wrong thing."

She sighed. "I hate that answer."

"So do I."

Across town, Lena sat with a reporter who'd asked for coffee instead of a call—an old trick to see if someone would talk more when the recorder stayed in a bag.

"I keep hearing the same phrase," the reporter said, stirring her cup. "*No immediate risk*. Who came up with it?"

Lena didn't flinch. "No one. It's descriptive."

"It's also everywhere."

"Because it's accurate."

The reporter leaned back. "Accurate can still be coordinated."

"Coordination implies intent," Lena said. "This is convergence."

The reporter raised an eyebrow. "You're good."

Lena met her gaze. "You are too. That's why this is frustrating."

"Help me," the reporter said. "What's the story *not* allowed to be?"

Lena chose a careful truth. "One that implies inevitability."

"Why?"

"Because inevitability demands action," Lena said. "And action demands certainty."

The reporter scribbled a note. "And you don't have certainty."

"No," Lena said. "We have alignment without closure."

They sat with that. Steam rose. Cups emptied.

"So," the reporter said finally, "if I run a piece about alignment without closure—"

"—your editor will ask what people should do," Lena finished.

"And I won't have an answer."

"You'll have a hundred answers," Lena said. "None of them usable."

The reporter closed her notebook. "That's the problem."

"Or the truth," Lena said.

Later that week, a leak tried to be a spark.

A partial slide deck surfaced online—screenshots cropped tight, equations stripped of captions, a single graph highlighted in red. The caption beneath it read: *They've been downplaying this for years.*

It flared. It faded.

Jun watched the cycle play out in real time. "This is what happens when fragments escape without scaffolding," he said.

Ruth nodded. "Scaffolding looks like endorsement."

"And endorsement looks like liability," Paul added.

A notification pinged. Another request for comment.

Ruth typed a response that would be reused, reshaped, and misunderstood anyway:

The materials circulating are incomplete and lack context. Ongoing research continues to refine understanding. There is no immediate risk.

She paused, then deleted the last sentence.

Jun noticed. "Why?"

"Because repetition turns accuracy into suspicion," Ruth said. "Let someone else say it this time."

Someone did. An editor paraphrased it. A host summarized it. A newsletter softened it.

The effect was the same.

Accuracy diffused as it traveled.

Each retelling preserved the facts while dulling their edges.

No version was wrong enough to correct.

None was sharp enough to hold attention.

In a small investigative shop, a reporter tried a different tack. He called a half-dozen scientists who'd published adjacent work and asked a single question: "What would convince you this matters now?"

The answers were thoughtful. Divergent. Conditional.

He sent the draft to his editor.

The editor wrote back: *This reads like a debate club transcript.*

The reporter replied: *That's because it is.*

Find a hook, the editor wrote. *Or park it.*

The reporter parked it.

Rumors kept mutating. One claimed a secret panel. Another insisted on a classified memo that didn't exist. They spread easily in the absence of clear explanation and any formal account. Each collapsed under scrutiny, leaving behind a residue of distrust that attached itself to the wrong things.

Paul shook his head. "We're losing the middle."

Lena didn't argue. "The middle doesn't travel well."

Ruth looked at the wall clock. "Neither does nuance."

The next call came from a radio host known for long-form conversations—no chyrons, no clips, just an hour and a patient audience.

"I won't ask you to predict," the host said. "I'll ask you to explain."

Lena hesitated. "Explain what?"

"How we decide what 'now' means."

They went quiet together, the question hanging there like a bridge no one had finished building.

"If I do this," Lena said, "you'll air it?"

"I'll air it," the host said. "But it won't trend."

"That might be the point," Lena said.

After they hung up, Jun said, "That's a risk."

"Yes," Ruth agreed. "But it's a different one."

The episode aired. It was careful. Dense. Earnest.

It was shared by people who already cared.

And that was the pattern, again.

The media didn't miss the story because it was absent. It missed it because it arrived without a moment to receive it.

By the end of the quarter, Lena understood the quiet truth beneath the noise: the press wasn't waiting for proof.

It was waiting for permission to simplify.

Until then, investigations would circle, questions would sharpen, and leaks would flash and vanish—each a rehearsal for a story that still refused to be told.

Somewhere ahead, she knew, there would be an image. Or a date. Or a failure with a name.

Until then, the truth would keep doing what it had been doing all along.

Moving.

Publishing.

Accumulating.

And failing, again and again, to become news.

∝ ∝ ∝

The permission never arrived.

No one knew how to grant it without also conceding control.

A veteran editor said it most plainly during a background call with Ruth. No hedging. No politeness.

"Look," he said, "if I run this, I'm either telling people nothing's wrong or something's coming. I can't thread the needle you're asking for."

Ruth closed her eyes. "What if the truth is that both are false?"

He laughed once, dry. "Then it's not a story yet."

"And when is it?" she asked.

"When it changes behavior," he replied. "Yours or theirs."

They both knew what that meant.

In another newsroom, a standards editor flagged a pitch because it was *indefensible.*

No falsifiable claim, the margin note read.

No event window.

No corrective action.

The reporter pushed back.

Isn't that the point?

The editor responded within minutes.

The point of journalism is not to explain why nothing can be done.

That line circulated quietly among the group after someone forwarded it without comment.

Jun stared at it longer than he needed to. "They're not wrong," he said.

"No," Lena replied. "They're just optimized for a different phase of reality."

The phase where consequences had already announced themselves.

The calls kept coming anyway.

A late-night voicemail from a freelance reporter: "I'm hearing this is being slow-walked."

A text from a producer: "Off record—are we going to look stupid for missing this?"

An email with no signature: *You can't keep this quiet forever.*

Lena deleted that one without replying. It misunderstood what was happening.

No one was keeping it quiet.

They were letting it speak in a language that didn't carry.

One afternoon, she sat in on a meeting she wasn't technically required to attend—communications, legal, two people from policy who rarely spoke unless they had to.

The question on the table wasn't *what do we say?*

It was *what happens if someone finally listens?*

"If a major outlet runs with it," legal said, "we'll need to clarify scope immediately."

"Clarify how?" Lena asked.

"That nothing actionable is implied."

Jun exhaled. "Actionable by whom?"

Legal didn't answer. They didn't need to.

Paul spoke instead. "We're treating action like a hazard."

"It is," legal said. "When it's unbounded."

Ruth folded her hands. "And in the absence of action, what do you think the public hears?"

Legal hesitated. "That things are under control."

"And if they aren't?" Lena asked.

Legal met her gaze. "Then we'll have wished we'd acted sooner."

Silence.

Not disagreement. Recognition.

Afterward, in the hallway, Jun said quietly, "Everyone's waiting for someone else to cross a line."

"Yes," Ruth said. "Because the first one across becomes responsible for whatever follows."

That was the truth no one put in print.

Responsibility, in media terms, didn't attach to accuracy. It attached to *activation.*

Who caused people to move.

To fear.

To demand.

To react.

As long as the story remained inert—true but inactive—it belonged to no one.

That was safety.

It minimized exposure without requiring denial. It allowed truth to persist without consequence.

Lena understood it fully now.

The media failed because it was functioning exactly as designed—rewarding clarity, urgency, and consequence, and filtering out everything that threatened to blur those lines.

The story had none of that.

It had no villain to confront.

No image to repeat.

No moment to replay.

No single voice to quote without qualification.

And worst of all, it had no instruction for what came next.

So it remained where slow truths often do—everywhere and nowhere. Discussed in rooms with time. Published in places with patience. Known by people trained not to shout.

Until something changed.

Until an event arrived that would retroactively give the story a beginning.

Lena sat at her desk late, scrolling through unanswered messages, thinking not about what the press would do when that moment came—but about how many chances there had been to tell the story *before* it demanded one shape.

The media wasn't asleep.

It was waiting.

And waiting, she knew now, wasn't neutral.

It was the quiet agreement to let reality make the first move.

∝ ∝ ∝

The pressure sharpened when someone decided to ask the question sideways.

It started with a voicemail Paul didn't return right away.

"Hey," the voice said—too casual, too practiced. "I'm not looking for confirmation. I'm looking for orientation. Call me."

The number belonged to a reporter Paul knew well enough to worry about.

When he called back, the reporter didn't bother with pleasantries.

"Off the record," she said immediately.

Paul sighed. "That phrase doesn't mean what it used to."

"Fine," she said. "Background. Deep background. Whatever makes you comfortable."

"That usually means nothing does."

She laughed once. "You're learning."

A beat.

"Here's what I'm hearing," she continued. "Everyone who knows the math is calm. Everyone who knows how institutions behave is nervous. That's not a coincidence."

Paul chose his words carefully. "People are cautious."

"No," she said. "They're *coordinated.*"

∝ ∝ ∝

Paul felt the familiar tightening in his chest. "That's an interpretation."

"It's a pattern," she replied. "And patterns are my job."

Silence stretched.

"Tell me this much," she said. "If I publish a piece saying the real story is that no one can say the real story yet—am I wrong?"

Paul closed his eyes. "You'd be incomplete."

"In which direction?"

"That's the problem," he said. "There isn't one."

She exhaled. "You realize how that sounds."

"Yes," Paul said. "That's why we haven't said it."

They hung up without resolution.

Elsewhere, Lena was dealing with a different kind of pressure.

A producer from an international outlet had requested an interview—not live, not taped, just "a conversation." The request came with a list of questions that looked innocuous until you read them together.

When did the language shift?

Who decides what counts as premature?

Is there a point at which caution becomes omission?

Lena read them twice.

"These are good questions," she said aloud.

Jun, standing in her doorway, nodded. "That's what makes them dangerous."

She agreed to the call.

The producer didn't waste time. "I'm not interested in accusing anyone," he said. "I'm interested in understanding the choreography."

"That implies intent," Lena replied.

"It implies learning," he said. "Which is what institutions do."

She paused. "What would you do with the answer?"

"I'd decide whether this is a story now or a story later," he said. "And if later—why."

Lena chose a narrow truth. "Because stories need edges. This doesn't have any yet."

"That sounds rehearsed," he said gently.

She smiled despite herself. "It's practiced."

"Then practice on me," he said. "Where does the pressure come from?"

Lena hesitated.

"From consequences," she said finally. "Not penalties. Consequences you can't undo once they're triggered."

The producer was quiet for a moment. "So you're managing behavior."

"We're managing interpretation," Lena said.

"And behavior follows," he replied.

"Yes," she admitted.

"That's a story," he said.

"Only if you can explain it without turning it into a scandal," she said.

He laughed softly. "That's never been the bar."

After the call, Lena felt exposed in a way she hadn't before. The questions were getting closer to the bone.

Jun found her staring at her screen.

"They're triangulating," he said.

"Yes," she replied. "And they're running out of patience."

The next leak wasn't technical.

It was procedural.

An internal email—harmless on its own—surfaced on a blog known for insinuation rather than evidence. The headline didn't claim wrongdoing. It didn't need to.

'No Immediate Risk' Becomes Internal Shorthand, Sources Say.

Paul read it twice, then forwarded it without comment.

Ruth replied a minute later. *This is how it starts.*

A reporter called Lena within the hour.

"I'm not publishing," he said quickly. "Not yet. But I need to ask."

"Ask," she said.

"Is 'no immediate risk' something you were told to say?"

"No," Lena replied. "It's something that turned out to be useful."

"Useful how?"

"It ended questions that couldn't be answered responsibly."

The reporter was quiet. "That's… not nothing."

"No," Lena said. "It isn't."

Another pause.

"So," he said, "who decides when it stops being useful?"

Lena looked out the window, at traffic moving with the confidence of people who assumed tomorrow would look like today.

"No one," she said. "That's the risk."

The reporter exhaled. "You realize I can't print that."

"I know," Lena said.

"But you also know I won't forget it."

"Yes," she replied.

By the end of the week, the tone had shifted.

Questions were sharper. Calls were shorter. Reporters stopped asking *what is happening* and started asking *who knows.*

Ruth summarized it during a late meeting.

"They're no longer looking for the story," she said. "They're looking for the moment it becomes undeniable."

"And until then?" Paul asked.

"Until then, they're applying pressure," Jun said. "Seeing what shakes loose."

Lena nodded. "They don't need proof. They need friction."

The media hadn't found the story.

But it had begun to sense resistance.

And resistance, unlike uncertainty, was something it knew exactly how to investigate.

∝ ∝ ∝

The Quiet Ethical Fracture

The fracture didn't announce itself as dissent.

It surfaced as discomfort.

In meetings, the arguments began to repeat with less conviction, losing their urgency even as they were restated, recycled from earlier sessions with only minor adjustments. People reached for the same phrases—*not mature enough*, *premature*, *needs more validation*—and felt, dimly, that they were no longer explanations so much as shields.

Lena felt it first in herself.

She noticed it when she re-read an email she'd drafted and realized she'd removed a sentence but because it would provoke a question she didn't want to answer yet.

She stared at the deleted line for a long time before closing the draft.

Across the hall, Jun was having a different conversation.

"This isn't about readiness," he said quietly, leaning against a filing cabinet while a colleague pretended to look for something inside it. "It's about tolerance."

The colleague frowned. "For what?"

"For uncertainty," Jun said. "For admitting we don't control how this lands."

"That's not our job," the colleague replied. "Our job is to make sure it *doesn't* land badly."

Jun tilted his head. "At what cost?"

The colleague hesitated. "At the cost of being early."

"That's not an answer," Jun said.

"It's a boundary."

Ruth encountered the fracture in a different form—during a closed session that had been scheduled as a formality and turned into something else.

A senior advisor spoke first. "Transparency is a duty," she said. "We're not here to curate reality."

Another responded immediately. "And we're not here to trigger panic with half-formed understanding."

"That assumes the public can't hold ambiguity," the first shot back.

"No," the second replied. "It assumes institutions can't survive being forced to act without options."

Ruth watched the exchange carefully.

Neither was wrong.

A third voice entered, weary. "We're arguing values without acknowledging constraints."

"And that's how ethical failures start," the first said. "When constraints replace principles."

"When principles ignore consequences," the second countered.

Ruth finally spoke. "Everyone here is acting rationally," she said. "That's the problem."

The room quieted.

"Rationality doesn't converge," she continued. "It branches. And once it does, the system has to choose which branch it treats as virtue."

Later, Paul found Lena in the break room, staring into a cup she hadn't touched.

"You look like you're losing an argument," he said.

"I think I already lost it," she replied.

"Which one?"

"The one where I tell myself we're being careful for the right reasons."

Paul leaned against the counter. "We are."

"Yes," Lena said. "And so is everyone who disagrees with us."

Paul frowned. "You think this splits?"

"I think it already has," she said. "We just haven't named the sides yet."

He was quiet for a moment. "What would you call them?"

She considered. "Those who believe transparency is a duty. Those who believe panic is a greater harm. And the rest of us—who believe the data isn't mature enough yet, and hope that belief keeps us from choosing."

Paul exhaled. "That's most people."

"Yes," she said. "That's what makes it dangerous."

The fracture widened not through confrontation, but through accumulation—small decisions stacking in different directions. One team pushed for clearer language. Another insisted on restraint. Review comments grew sharper. Emails took longer to answer.

No one defected.

No one blew a whistle.

They simply began to trust different futures.

Jun put it bluntly during a late discussion that ran past its scheduled end.

"At some point," he said, "we stop disagreeing about data and start disagreeing about responsibility."

"And when is that?" someone asked.

Jun didn't answer right away. "When silence starts to feel like action."

That line stayed with Lena long after the meeting ended.

She thought about how easily responsibility had slipped its definition. How containment had first felt like care. How care had turned into caution. How caution now bordered on avoidance.

None of it felt corrupt.

That was the most unsettling part.

Everyone could justify their position. Every argument rested on evidence, precedent, and fear of harm. No one was acting in bad faith.

The fracture wasn't between good and bad actors.

It was between incompatible versions of duty.

And as it deepened, Lena understood something she hadn't before:

The most dangerous ethical conflicts aren't the ones where someone chooses wrong.

They're the ones where everyone chooses *reasonably*—and the system fractures anyway, because reason alone cannot decide when truth becomes an obligation.

The break was quiet.

But it was real.

And once it formed, it wouldn't heal on its own.

And then the room learned what it meant to live with it.

∝ ∝ ∝

The conversations grew more careful.

Not calmer—*careful.*

People started qualifying agreement before they voiced it, as if testing which version of responsibility the room would tolerate.

"I don't disagree," someone would say.

"But—"

The *but* did more work than the rest of the sentence.

In one meeting, a senior analyst finally pushed back hard enough to make the air change.

"We keep saying the data isn't mature," she said, palms flat on the table. "At what point does that stop being a scientific assessment and start being a moral cover?"

A murmur moved through the room. Not outrage. Recognition.

Paul answered first. "When maturity becomes a pretext."

"And who decides that?" she asked.

Paul hesitated. "Consensus."

She shook her head. "Consensus is what you get *after* people are allowed to disagree publicly."

Jun leaned in. "Or what you get when disagreement becomes too expensive."

That landed harder than anyone expected.

Ruth intervened. "Let's slow this down. No one here is arguing for permanent silence."

"No," the analyst said. "We're arguing for indefinite postponement. Which is functionally the same thing."

"That's unfair," someone else said. "We're talking about harm reduction."

"And so am I," she replied. "I just think harm includes infantilizing the public."

A pause.

"That assumes the public wants ambiguity," Paul said.

"No," she countered. "It assumes they deserve honesty about it."

Jun watched the exchange closely. He could see the fault line now—cleaner than before.

"This isn't about data," he said. "It's about whether uncertainty belongs only to experts."

Ruth met his eyes. "Or whether experts have an obligation to prevent uncertainty from becoming weaponized."

"By whom?" Jun asked.

"Anyone," she said. "Including us."

Later, in a smaller room, the tone changed again—less formal, more dangerous.

A colleague closed the door behind them. "Can I say something without it becoming a position?"

"That depends," Lena said. "Do you want it remembered?"

The colleague laughed nervously. "I'm starting to think transparency is being treated like a luxury good."

Jun frowned. "Explain."

"It's fine when nothing's at stake," the colleague said. "But once stakes appear, transparency becomes optional—something you do *after* you've secured outcomes."

Paul crossed his arms. "That's cynical."

"Is it?" the colleague asked. "Or is it just how institutions survive?"

Lena felt the weight of that question settle into her chest.

Another voice, quieter. "What scares me," someone said, "is that if we're wrong, we'll be blamed for hiding it. But if we're right, we'll be blamed for not preventing what follows."

Silence.

Ruth finally spoke. "That's the bind. There is no path that avoids blame."

"So we choose the one that minimizes regret," Paul said.

Jun shook his head. "No—we choose the one that minimizes *visibility* of regret."

Paul looked at him sharply. "You think that's what we're doing?"

"I think it's what we're drifting toward," Jun said. "And drift is still movement."

The fracture widened another fraction.

Not enough to break anything.

Enough to change how people looked at each other.

Emails grew longer, more careful. Meetings ended without clear resolutions. People began looping legal into conversations that had once been purely scientific—not to block action, but to *anticipate consequence.*

Lena noticed something else too.

People started confiding selectively.

Not based on rank.

Based on values.

Those who leaned toward transparency found each other quietly. Those who feared panic did the same. Conversations happened in pairs, then trios—never openly coordinated, but unmistakably aligned.

The system wasn't splitting.

It was *sorting.*

Jun named it one night, standing with Lena by the windows as the building emptied.

"We're forming camps," he said.

She nodded. "Without admitting it."

"That's the dangerous phase," he said. "When disagreement becomes identity."

Lena watched the lights go out floor by floor. "Do you think this ends in a break?"

Jun considered it. "I think it ends in a moment when someone is forced to act in a way that makes neutrality impossible."

"And who do you think that will be?" she asked.

He didn't answer.

Because they both knew the truth: ethical fractures don't announce who will carry them.

They wait.

They deepen quietly.

And when they finally surface, it's never because someone wanted to be brave.

It's because staying silent stopped feeling like the safer choice.

The moment neutrality failed came dressed as a procedural question.

It was raised near the end of a meeting that had already gone long, when people were tired enough to stop guarding their instincts.

"Should we add a disclaimer?" someone asked. "Just a line. For clarity."

"What kind of disclaimer?" Ruth said.

The speaker hesitated. "Something that acknowledges downstream implications without… emphasizing them."

Jun looked up. "That's not a disclaimer. That's a hedge."

Paul leaned back. "Hedges exist for a reason."

"Yes," Jun said. "To avoid commitment."

The room went quiet again. Not the calm kind—the kind where everyone feels the floor shift a few centimeters.

Lena spoke carefully. "What problem is the disclaimer solving?"

"Exposure," legal said from the corner. "If someone claims they weren't informed—"

"Informed of what?" Lena asked.

Legal paused. "Of the possibility."

Jun nodded. "So we're not talking about public understanding. We're talking about liability."

"That's not cynical," legal replied. "That's responsible."

Ruth interjected. "Responsibility to whom?"

"To the institution," legal said, without apology.

That was the first time someone said it that plainly.

A few people exchanged glances. No one objected.

Paul broke the silence. "Institutions don't exist in a vacuum. Protecting them protects the work."

"And when those interests diverge?" Jun asked.

"They don't," Paul said. "Not if we do our jobs."

Jun shook his head. "That's faith, not logic."

Later, in the hallway, Lena caught up to Ruth.

"That landed harder than you expected," Lena said.

Ruth nodded. "Because it stripped away the fiction."

"Which one?"

"That this is still just about data."

They walked in silence for a few steps.

"Do you remember when transparency felt like default?" Lena asked.

"Yes," Ruth said. "When consequences were abstract."

"And now?"

"Now transparency is a decision," Ruth replied. "Which makes it political whether we like it or not."

That night, an email chain split in two.

It started innocently—comments on wording—but within hours it had forked into parallel threads, each with a different tone, a different set of recipients.

One emphasized caution.

The other emphasized obligation.

No one accused anyone of bad faith. They didn't need to.

Jun replied to one thread with a single sentence:

We should be honest about why we're hesitating.

The response came back from Paul, privately.

Honesty without restraint is not a virtue.

Jun typed, deleted, typed again.

Restraint without honesty is not ethics.

Paul didn't reply.

The next day, a junior researcher knocked on Lena's door.

"I don't know who to talk to," she said, voice low. "But I need to ask."

Lena gestured for her to sit.

"If we believe this matters," the researcher said, "and we also believe people can't handle it yet—how do we decide when 'yet' ends?"

Lena didn't answer right away.

"We don't," she said finally. "That's the part no one wants to admit."

The researcher frowned. "Then what stops this from becoming permanent?"

Lena met her gaze. "People like you asking that question."

The researcher absorbed that, then nodded slowly. "That's not very comforting."

"No," Lena agreed. "But it's true."

As the days passed, the fracture became visible in small betrayals of routine.

Someone cc'd an extra person "just in case."
Someone else stopped forwarding a draft to the full group.
Meetings were scheduled with tighter guest lists.

Nothing dramatic.

Just enough to signal trust was no longer universal.

Jun noticed it most clearly when a long-standing collaboration stalled. No one wanted to be the one to escalate.

"This is how systems freeze," he said to Lena. "Not from conflict. From fear of being the first mover."

"And then?" she asked.

"And then someone moves anyway," Jun said. "Not because they're reckless—but because they can't stand the quiet anymore."

Lena thought of the journalist questions. Of the private draft on her laptop. Of the disclaimer debate that had cracked something open.

She understood now that the ethical fracture wasn't about transparency versus panic.

It was about *agency.*

Who had the right to decide when silence stopped being care and started being control.

And as that question spread—unanswered, unavoidable—the system did what it always did under unresolved moral load.

It redistributed the burden. Quietly.

Until it found someone willing, or desperate enough, to carry it alone.

That's how fractures become policy.

∝ ∝ ∝

The burden landed first on people who hadn't asked for it.

A mid-level program manager sent an email that was meant to be neutral—an update, nothing more. She included a sentence that acknowledged downstream implications explicitly, then flagged it as *contextual* rather than *conclusive*. It was careful. Responsible. Boring.

Within an hour, her phone rang.

"I'm not saying it's wrong," her supervisor said. "I'm saying it changes how this reads."

"How?" she asked.

"It invites interpretation."

"That's what context does," she replied.

A pause. "Let's talk about removing it."

She hesitated. "Because it's inaccurate?"

"No," he said. "Because it's unnecessary."

She closed her eyes. "Unnecessary to whom?"

Another pause. Longer this time. "To the purpose of this release."

When the call ended, she stared at the sentence on her screen. It wasn't dramatic. It didn't accuse. It simply connected two things that had been kept adjacent but separate.

She deleted it.

The email went out. No one complained. No one thanked her.

The system absorbed the correction and moved on.

When Jun heard about it later, he didn't react the way Lena expected.

"That's how it spreads," he said quietly. "Not through leaders. Through normalization."

"What would you have done?" Lena asked.

He considered it. "I don't know. That's the point."

Another conversation, different day.

Paul sat with legal and communications, the tone clipped, efficient.

"We need to be consistent," communications said. "Inconsistency reads as uncertainty."

"We *are* uncertain," Paul replied.

"Yes," legal said. "But we can't be *visibly* uncertain."

Paul leaned back. "So we present confidence we don't have?"

"Professional confidence," communications said. "In process, not outcome."

Paul nodded slowly. "That's a distinction."

"It's a shield," Jun said from the doorway. He hadn't been invited. He hadn't asked.

Communications bristled. "We're managing expectations."

Jun met their gaze. "You're managing responsibility."

Legal closed their folder. "Responsibility to prevent harm."

Jun didn't blink. "Responsibility to decide who gets to assess it."

The room went still.

Paul exhaled. "We're not doing this now."

Jun stepped back. "That's the problem. We never are."

Later, Ruth convened a smaller meeting. No agenda. No minutes.

"We need to acknowledge what's happening," she said. "Not to the public. To ourselves."

A few nods. No objections.

"There are three positions in this building," she continued. "Those who believe transparency is a duty regardless of consequence. Those who believe preventing panic is a higher duty. And those who believe the data isn't mature enough to force either choice."

She paused. "The third group is shrinking."

Lena felt that land.

Ruth went on. "I'm not asking anyone to switch sides. I'm asking us to recognize that neutrality is becoming performative."

A voice from the back spoke up. "Are you saying we have to choose?"

Ruth shook her head. "I'm saying choices are being made whether we name them or not."

Jun spoke softly. "And unnamed choices are the ones no one takes responsibility for."

Paul looked at him. "You keep saying that like it's an accusation."

"It's a warning," Jun replied.

That night, Lena received an anonymous message through a channel she hadn't used in years.

How long do you think you can hold this line?

She didn't respond.

The question assumed there was still a single line to hold.

The fracture had done its work. It had turned collective judgment into private calculus. People were no longer asking *what is right?* They were asking *what can I live with?*

That was the quiet danger.

Ethics, once internalized, stop sounding like rules. They sound like instincts. And instincts, Lena knew, were shaped by repetition, by reward, by fear of isolation.

By the time she shut down her computer, the building mostly dark, she understood what the fracture really meant.

Not that the institution would fail.

But that when the moment came—when silence finally tipped into harm—no one would be able to say, honestly, that they had chosen it together.

They would say it happened.

And that, she thought, was the most dangerous story of all.

And "happened" was the safest verb they had left.

∝ ∝ ∝

The fracture stopped being theoretical the day someone asked for a name.

It came during a meeting that had already lost its original purpose. Slides were still on the screen, untouched for more than ten minutes.

People were talking around them now, not to them.

"If this goes sideways," a policy director said carefully, "who owns the call we made?"

No one answered.

She pressed. "I'm not asking rhetorically."

Paul shifted. "Ownership is distributed."

"That's not how blame works," she replied. "Or history."

Jun leaned forward. "That's why distributed decisions feel safe while they're happening."

"And dangerous afterward," Lena added.

The policy director nodded. "Exactly. So I'm asking again—who is responsible for the posture we've taken?"

Ruth spoke, voice steady. "We are."

The room stilled.

"All of us?" someone asked.

"Yes," Ruth said. "Which means none of us can pretend neutrality anymore."

A few people exchanged looks. One person glanced down at their phone, then put it face down on the table as if that gesture itself mattered.

"That's not realistic," legal said finally. "Responsibility has to attach to authority."

Jun smiled faintly. "Authority has been very comfortable letting responsibility blur."

Legal bristled. "We're trying to prevent harm."

"And so is everyone who disagrees with you," Jun said. "That's the fracture."

Another voice—quieter, from someone who rarely spoke—cut in. "I don't think I believe we're preventing harm anymore."

Heads turned.

The speaker swallowed. "I think we're preventing *reaction.* And we've started calling that the same thing."

Silence followed. Heavy. Unavoidable.

No one reached for process this time.

The framework they relied on had been named too clearly.

Paul broke it. "Reaction can be harm."

"Yes," the speaker said. "But so can delayed agency."

Lena felt the room tilt—not toward chaos, but toward honesty.

Ruth nodded slowly. "Say more."

The speaker hesitated, then continued. "Every day we wait, we're deciding for other people that they don't get to weigh uncertainty yet. We tell ourselves it's temporary. But we've stopped defining the conditions under which it ends."

Jun exhaled. "That's what I've been trying to say."

Paul rubbed his face. "So what's the alternative? We dump complexity into public discourse and hope for the best?"

"No," Lena said. "We stop pretending that caution absolves us from choice."

The policy director leaned back. "Meaning?"

"Meaning," Lena said, "we acknowledge—internally, at least—that we're choosing stability over participation. And we accept the ethical cost of that choice instead of hiding it inside process."

Legal shook their head. "That's not how institutions survive."

Ruth met their gaze. "Institutions don't survive by avoiding ethics. They survive by confronting them early enough to adapt."

A long pause.

Someone laughed quietly—not in humor, but disbelief. "Listen to us. We sound like a seminar."

Jun replied, "Seminars end. This won't."

After the meeting broke, people didn't linger the way they used to. Conversations split off in pairs, then ended abruptly. No one wanted to be overheard saying the wrong thing to the wrong person.

In the hallway, Paul caught up to Lena.

"You realize," he said, "that if we start naming this as an ethical choice, we invite scrutiny we can't control."

"Yes," Lena replied.

"And if we don't?"

She stopped walking. "Then we become the kind of institution that mistakes silence for neutrality."

Paul looked at her for a long moment. "You think that's worse."

"I think," she said carefully, "it's harder to defend."

He nodded once. "History's cruel that way."

That evening, Jun sent a message to a small group—nothing inflammatory, just a question framed as an invitation:

At what point does delaying truth become deciding on behalf of others?

No subject line. No recommendation.

Replies trickled in. Some cautious. Some grateful. Some angry.

None indifferent.

Lena watched the thread grow and felt the shape of the fracture clearly now—not a crack that would shatter the institution, but a seam that could no longer be ignored.

Everyone was acting rationally.

Each position made sense on its own. Each was defensible. Each was grounded in evidence, experience, and fear of harm.

That was what made it dangerous.

Because when reason diverges without a shared ethical anchor, systems don't fail loudly.

They fracture quietly—until one day the silence itself becomes the decision no one remembers making.

The system didn't choose a side. It chose a schedule.

∝ ∝ ∝

The fracture didn't resolve.

It stabilized.

That was the signal no one wanted to name.

It happened first in scheduling, where urgency could be softened without anyone having to quietly justify the change. A meeting that had originally been marked *urgent* was reclassified as *standing.* Another was pushed to biweekly "to allow for synthesis." The calendar filled, not with alarms, but with placeholders.

Ruth noticed it when an assistant asked what to call the next session.

"Is this exploratory?" the assistant asked. "Or governance?"

Ruth hesitated. "Governance," she said finally.

The word mattered more than it should have.

Governance meant structure.

Structure meant continuity.

Continuity meant the problem was no longer exceptional.

Jun flagged it in an email that went unanswered.

Once we formalize this, we stop treating it as unresolved.

Paul replied an hour later.

Formalization is how we keep unresolved things from metastasizing.

Jun didn't argue. He didn't need to. The machinery was already turning.

A document appeared—nothing dramatic. Just a shared folder with a neutral title and a clean taxonomy. Subfolders populated themselves: background, literature, communications considerations, risk framing. Someone added version control.

Lena opened it and felt the shift immediately.

This wasn't avoidance.

It was incorporation.

At the next meeting, legal spoke with visible relief. "We're setting up a review cycle," they said. "Quarterly reassessment. That gives us defensibility."

"And a working group," communications added. "Small. Cross-functional."

Paul nodded. "We'll need a budget line if this is ongoing."

Ruth caught Lena's eye across the table. Neither of them spoke.

Jun did. "You realize what this does."

Paul looked at him. "It gives us process."

"It gives the problem a home," Jun said. "And once it has a home, it stops knocking."

No one contradicted him.

Because the truth was visible now, undeniable in its banality.

The issue had become administratively normalized.

It had a file.

A review cycle.

A working group.

A budget line.

A reassessment date.

Once that happened, urgency died.

The system had absorbed the problem.

PHASE 4 — THE PUBLIC TRANSLATION GAP

The first signs didn't arrive as warnings.

They arrived as trivia.

That was the first translation.

A teacher mentioned, offhand, that the Moon was drifting away a few centimeters a year—something she'd learned decades ago, filed under *interesting but irrelevant.* Her students nodded, wrote it down, forgot it by the next quiz.

A magazine ran a sidebar in the 1970s about laser reflectors left behind by astronauts. The tone was celebratory. Precision science. Proof of ingenuity. The fact that the measurements revealed motion—slow, relentless—was treated as color, not content.

Later, a documentary used the same fact as a visual flourish. A diagram. An animation. The narrator's voice warm, reassuring.

"The Moon," he said, "is still doing what it has always done."

No one objected.

Reassurance did the work of judgment.

The fragments leaked the way all technical truths do—through simplification.

A newspaper headline in the 1980s read *Moon Slowly Drifting Away from Earth.* Slowly did all the work. Readers skimmed, shrugged, and turned the page.

In the 1990s, a science show joked about longer days in the far future. "Don't worry," the host laughed, "you won't notice."

The audience laughed with him.

The signs were there, but they arrived pre-digested. Stripped of implication. Packaged as curiosity. Each appearance minimized the next by repetition. If it had been important, surely someone would have said so differently.

Someone did.

A researcher gave a public talk in the early 2000s and tried—carefully—to explain tidal coupling as a system, not a phenomenon. He lost the room halfway through. The questions afterward were polite and off-target.

"So… is this dangerous?" someone asked.

"No," he replied, truthfully. "Not immediately."

That was the sentence that survived.

Survival wasn't the same as meaning.

A blogger summarized his talk the next day: *Scientist Says No Immediate Risk from Moon's Motion.* The comments filled with jokes about werewolves and cheese. Someone posted a meme.

The signal drowned in noise.

Over time, the fragments accumulated without cohering.

A retired engineer remembered being taught that Earth's rotation was slowing—milliseconds per century. He'd filed it next to continental drift and stellar lifetimes: facts too slow to matter.

A climate skeptic stumbled onto lunar recession numbers and dismissed them as another long-term abstraction. "Come back in a million years," he wrote. "Then we'll talk."

A college student watched a video explaining leap seconds and thought it was clever that clocks needed adjustment. She never asked why.

None of them were wrong.

They were just incomplete.

Incompleteness was functional.

The public didn't reject the facts.

They absorbed them in pieces, each fragment translated into something manageable—interesting, ignorable, funny, or harmless. Without synthesis, meaning collapsed under familiarity.

And then, quietly, the timeline crossed a threshold no one had marked.

The past sixty years showed consistency. Measurement refined. Precision increased. Models sharpened. The curve remained smooth enough to reassure.

The next hundred years didn't offer drama either—only extrapolation.

Centimeters per year.

Milliseconds per century.

Small numbers, stretched thin.

Presented plainly, they sounded like nothing.

Presented together, they sounded like inevitability.

Inevitability does not demand attention.

But inevitability had no place to land in public discourse. It required patience without payoff, attention without spectacle. It asked people to care about change that refused to perform.

So when fragments escaped—through articles, lectures, interviews—they arrived misinterpreted, minimized, mocked, or stripped of meaning entirely.

Because translation is loss.

And the loss accumulated at every boundary between expert language and human attention.

This was the gap.

Not between truth and falsehood—but between signal and reception.

Not heroes discovering secrets—but ordinary people receiving noise, each trying to fit it into lives already crowded with urgency.

By the time the public entered the story, they weren't missing facts.

They were missing *coherence.*

Coherence is not self-assembling.

And that absence—quiet, rational, understandable—was the most consequential condition of all.

Because once meaning fractures at scale, no amount of accuracy can reassemble it quickly.

The system had spoken.

The public had heard *something.*

And now, for the first time, the story belonged to people who didn't know what they were listening for—only that the sound had been there all along.

∝ ∝ ∝

The gap widened the moment the fragments began to collide with lived experience.

Not catastrophes.

Inconveniences.

A navigation system that required a correction no one could explain simply.

A timing anomaly in satellite coordination that engineers quietly compensated for and never named.

A footnote in an infrastructure report noting that long-term synchronization assumptions would need revision "in future frameworks."

Nothing broke.

Things just needed *adjustment* more often than before.

People noticed in ways they couldn't quite articulate.

A radio host mentioned, jokingly, that leap seconds were being retired because "time itself was getting annoying." Listeners laughed. One called in to ask why that was happening now and not before.

"Well," the host said, "turns out Earth's a little sloppy."

The phrase stuck for a week. Then vanished.

A popular science podcast revisited lunar laser ranging for an anniversary episode. The hosts marveled at the precision—centimeters measured across hundreds of thousands of kilometers. One of them paused.

"So… we've been watching this drift for decades," she said. "Are we… supposed to do anything with that?"

There was a beat of silence.

"No," the other replied. "It's just cool."

They moved on.

In comment sections and forums, the fragments began to attract interpretation rather than understanding.

Someone stitched together charts from different decades and posted them with the caption: *Why isn't anyone talking about this?*
Replies ranged from patient debunking to outright ridicule.

"It's normal."
"It's always been like that."
"You're misunderstanding the scale."
"Come back in a million years."

All of them true. None of them sufficient.

A high school teacher tried to explain tidal forces to a class already exhausted by climate anxiety, economic uncertainty, and endless alerts. Halfway through, a student raised her hand.

"So is this another thing we're supposed to worry about?"

The teacher hesitated.

"No," he said. "Not worry. Just… know."

The student nodded, unsatisfied, and went back to her phone.

That was how the signs passed through the public—filtered by fatigue, humor, dismissal, and the quiet triage people perform every day just to function. Facts that arrived without a role to play were stored wherever the mind kept trivia.

The past sixty years, seen this way, were not silent.

They were noisy with fragments.

The next hundred would be louder—because the systems people relied on would grow more sensitive to small drifts. Precision would increase. Margins would thin. Adjustments would stack.

None of this would feel like a warning.

It would feel like friction.

Friction is how systems admit strain without declaring failure.

This was where the human stories belonged—not in discovery, not in denial, but in reception, in the quiet space between awareness and response, with no demands for action. In the way people encountered information already stripped of urgency and tried, unsuccessfully, to decide whether it mattered to their lives.

∝ ∝ ∝

SIGNS FROM THE PAST SIXTY YEARS

1969: The measurement era becomes permanent

The Moon changed the day light came back.

Not symbolically. Literally.

Until then, distance had been something you *estimated*. You triangulated. You averaged. You inferred. Astronomy was a discipline of approximation, of patience, of accepting that most scales were too large to touch directly. The Moon's orbit could be calculated to impressive fidelity, but it still lived in the realm of models—curves sketched from observation, refined by theory, corrected by assumption.

Then a pulse of light left Earth.

It wasn't dramatic. No one watched it go. There was no visible beam cutting through space. A laser fired, photons dispersed, and most of them vanished into nothing. But a fraction struck something new on the lunar surface: a small array of corner-cube reflectors, passive and unpowered, bolted into the regolith like an afterthought.

The light came back.

Not metaphorically. It returned after a delay that could be measured, argued over, and measured again. The round-trip time—there and back—collapsed an astronomical distance into a number that fit on a lab printout, small enough to be handled. The Moon stopped being "far" in the abstract sense and became far in milliseconds.

Time replaced distance.

That exchange never reversed.

That substitution mattered more than anyone said out loud.

Because it changed what counted as knowing.

For the first time, the space between Earth and Moon was no longer inferred from orbital mechanics alone. It was measured directly, repeatedly, with an accuracy that belonged to metrology, not stargazing. The same kind of precision used to calibrate instruments on Earth was suddenly being applied across hundreds of thousands of kilometers of vacuum.

When the first results arrived—centimeter-level accuracy across a gap larger than any terrestrial scale—the reaction was immediate and celebratory. It was framed as a technical triumph. Proof that the reflectors worked. Proof that the lasers were strong enough. Proof that the math closed.

We had done something clever.

Cleverness delayed consequence.

Papers were written. Talks were given. The achievement was folded neatly into the larger narrative of Apollo: flags, footprints, rocks, ingenuity. The reflector arrays were props in that story, absolute evidence enough that the Moon landings had not only reached a destination but left behind infrastructure.

What went largely unremarked was that the experiment didn't end.

Permanence arrived quietly.

It couldn't.

Unlike samples or photographs, this was a measurement that begged to be repeated. The reflector would still be there tomorrow. The laser would fire again. Another interval would be recorded. Another number added to the sequence.

One data point invites curiosity.
Two invite comparison.
Three invite expectation.

And expectation invites continuation.

Quietly, a permanent relationship was established. Earth began checking in on the Moon—not to see what it looked like, but to see where it *was*. And then, inevitably, to see whether that position changed.

The Moon, once a backdrop, became a clock.

Clocks demand continuity.

Each pulse of light asked the same question: *how far now?* Each answer differed by an amount too small to feel or to dramatize, but impossible to ignore once you were looking for it. That was the irreversible step.

Irreversibility does not announce itself.

We were no longer observing the Moon as an object in the sky.

We were timing light as it interacted with a system.

And once you start timing something—once you commit to precision, once you replace awe with intervals—you don't stop. Because stopping would mean choosing not to know the next number. And after the first measurement came back clean, there was no scientific, technical, or cultural reason to make that choice.

So the pulses continued.

Night after night.

Year after year.

The Moon changed the day light came back—it entered a regime where even the smallest drift could no longer hide inside approximation.

From that moment on, the question was never whether anything was happening.

It was how long it would take before the accumulation of numbers meant something we could no longer treat as trivia.

∝ ∝ ∝

1970s–1990s: The quiet accumulation

More reflectors followed.

Apollo 14 placed another array in a different location, angled differently, designed to answer a quiet question no one put in a press release: *was the result real, or was it luck?* Apollo 15 added a larger one, heavier, more efficient, capable of returning stronger signals. The Soviets left their own reflectors behind on robotic rovers—different engineering traditions, assumptions, but the same intent.

Each new reflector removed an excuse.

Different latitude.
Different longitude.
Different hardware.
Same answer.

Redundancy crept in the way it always does in mature science: cautiously at first, then with confidence. One mirror might be a fluke. Two might be coincidence. Three demanded attention. Four made denial embarrassing.

The Moon didn't care.

The signal came back clean from every direction. Not dramatically clean—just consistently so. The numbers lined up. The corrections agreed. The residuals shrank. Error bars behaved.

Year after year, measurement after measurement, the curve held its shape.

It didn't spike.
It didn't wobble.
It didn't misbehave in ways that forced explanation.

It was smooth.

That smoothness did more than reassure. It anesthetized.

Anesthesia is a form of success.

In the early years, people still leaned forward when the plots appeared. They asked careful questions. They checked the math. They argued about uncertainties. But as the decades passed and nothing unexpected happened, attention shifted elsewhere.

Consistency stopped being evidence and became background.

Background is where meaning thins.

The curve was no longer something you *studied.* It was something you *used.*

Boredom, in science, is success.

It means the phenomenon has stopped demanding explanation and started functioning as infrastructure. No one debates the voltage in a wall socket unless it fluctuates. No one questions gravity unless it fails. The Moon's distance, once a subject of wonder, became a parameter.

You corrected for it.

Orbital models included it automatically. Navigation algorithms accounted for it without comment. Papers cited it in passing, then moved on to what they considered the real work.

The charts accumulated quietly in databases and appendices, each new line extending the same gentle slope. Graduate students learned to plot them as an exercise, not a revelation. Senior researchers stopped including the figures in talks unless someone asked.

No one felt the need to narrate something that refused to change character.

Nothing dramatic happened.

That wasn't a disappointment.

That was confirmation.

Confirmation closed the question without ending it.

Dramatic data demands stories. Stable data dissolves into assumptions. Over time, the Moon's recession ceased to be something you noticed at all. It became a correction factor—a number you subtracted so the rest of your calculations would behave.

The absence of surprise was the strongest signal of all.

Because in a system that is wrong, something eventually breaks. Noise appears. Deviations force revision. But here, revision never arrived. Only refinement. Smaller uncertainties. Cleaner fits. Better confidence.

The charts didn't spike because nothing spiked.

They didn't wobble because nothing wobbled.

They drifted.

And drift, when it is smooth enough, does not feel like motion. It feels like permanence.

∝ ∝ ∝

2000s: The recession rate becomes a one-liner

Eventually, the result became simple enough to teach.

That was the threshold.

Teachability replaced urgency.

Not when it was discovered, not when it was confirmed, but when it could be reduced to a sentence short enough to fit on a classroom slide without footnotes.

The language changed first.

Early explanations had been cautious, wrapped in conditional phrasing and careful framing. Over time, as decades of data piled up without contradiction, the hedging fell away. Educational material stopped qualifying and began stating the conclusion plainly: the Moon is moving away from Earth at roughly 3.8 centimeters per year.

No drama. No caveats bolded for emphasis. Just a number.

The cause was presented as tidy and almost soothing. Tidal friction. Angular momentum. A slow exchange of energy between Earth's rotation and the Moon's orbit. You could draw it on a whiteboard with arrows and curved lines. You could explain it with a calm voice and a few analogies about ice skaters and spinning stools.

Nothing about it sounded alarming.

And it wasn't, framed that way.

Three point eight centimeters a year is small enough to disappear into intuition. It sits comfortably beside other geological and astronomical facts people are trained to ignore for their own sanity. Continental drift. Stellar evolution. The cooling of the Sun. Processes that belong to textbooks, not calendars.

It is a scale designed to reassure.

A fact that carries no demand. No implication that anyone should do anything differently tomorrow, next year, or even within a lifetime. A number that answers curiosity without inviting responsibility.

The sentence began appearing everywhere.

In NASA outreach material meant to inspire rather than unsettle. In middle-school textbooks as a boxed sidebar. In documentaries, delivered with a smile and followed immediately by reassurance: *don't worry—you'll never notice.* The figure was repeated so often, so consistently, that it acquired the dull familiarity of a household statistic.

How far away is the Moon?

About three hundred eighty thousand kilometers.

How fast is it moving away?

About three point eight centimeters a year.

Call and response. Question and answer. Closed loop.

Nowhere did it land as a warning, because nothing about its presentation asked to be treated that way. It wasn't framed as part of a system. It wasn't connected to timekeeping, orbital dynamics, or long-term coupling. It was offered as an isolated curiosity, a marvel of measurement rather than a property of motion.

The number became safe through repetition.

People learned it the way they learned the speed of light or the distance to the Sun—facts impressive enough to admire, abstract enough to ignore. If it mattered in any practical sense, surely someone would have said so differently.

Someone didn't need to.

The truth was never hidden.

It was never classified.

It was never denied.

It was simply packaged at a scale that drained it of consequence.

Filed under trivia.

Trivia is how facts survive without consequence.

∝ ∝ ∝

2010s–2020s: Precision becomes ordinary

By the time lunar laser ranging surfaced again in public conversation, it arrived stripped of any sense of discovery.

The question was no longer *can we measure this?*
That question had been settled long ago.

What remained was *how precisely*.

The framing shifted accordingly. Press releases, outreach pieces, and institutional summaries stopped treating the experiment as an exploration of the Moon and began presenting it as a demonstration of capability. The emphasis moved from celestial mechanics to engineering prowess. The story was no longer about what the Moon was doing, but about what we could do to it—how accurately we could reach across space and touch something that never moved closer.

Agencies spoke casually about hitting reflectors the size of luggage at lunar distance, as if threading a needle across hundreds of thousands of kilometers were an expected outcome rather than an extraordinary one. The language was relaxed, confident, almost bored. What would once have sounded miraculous was now described as routine.

And in a way, it was.

Centimeter accuracy became millimeter refinement. Error bars tightened. Corrections grew smaller. The excitement lived in the decimals, not the phenomenon itself. Researchers spoke with pride about shaving uncertainties down, about noise reduction, about the elegance of a clean return signal.

The experiment expanded sideways.

Lunar laser ranging became a platform rather than a question. It was used to test aspects of general relativity, to probe the Moon's internal structure, to constrain models of Earth–Moon dynamics. Papers multiplied. Applications diversified. The original purpose—tracking distance over time—was absorbed into a larger ecosystem of precision science.

The Moon became a proving ground.

A place where theory could be stress-tested, instruments calibrated, and assumptions confirmed. The data stream was no longer remarkable for what it revealed about change, but for how stable and reliable it was as a reference.

We celebrated the tool.

We celebrated the ability to aim, to fire, to receive. We marveled at the choreography required to send a pulse of light into space and catch its faint echo as it returned. We told the story as one of mastery, of technological maturity, of control.

And in doing so, we quietly inverted the meaning of the measurement.

The experiment was no longer asking *what is happening?* It was proving *how well we can ask*.

The numbers kept arriving, as they always had. Each one agreed with the last within increasingly narrow margins, with no abrupt deviations and consistent with long-term expectations. The drift remained smooth. The curve remained obedient. The signal said the same thing it had been saying since the first return.

But that message was no longer foregrounded.

It was background noise beneath the triumph of precision.

We could hit a suitcase on the Moon with a pulse of light. We could do it repeatedly, predictably, almost casually.

And still, we couldn't make the meaning travel.

Because precision without synthesis does not create understanding. It creates confidence. And confidence, once achieved, has a way of convincing systems that the question has already been answered—even when the answer has been quietly accumulating, unchanged, for decades.

A second sign people actually touched: timekeeping

The Moon doesn't only move away.

It pulls.

That pull is constant, indifferent, and patient. It drags against Earth's oceans, raising tides that slide across coastlines and retreat again. In doing so, it trades energy with the planet itself—stealing a trace of rotational momentum and giving it back as orbital distance. The exchange is small in any given moment, but relentless across time.

Earth responds the only way it can.

It slows.

Not in a way bodies can feel. Not in a way days suddenly stretch into something unfamiliar. The change is measured in milliseconds per century—fractions so small they vanish into lived experience. No sunrise arrives noticeably late. No afternoon lingers long enough to be remarked upon.

But systems notice.

Atomic clocks don't average. They count. They expect the world to keep pace with numbers that never drift. Against that standard, even a millisecond matters. Especially when systems stack—navigation, communication, finance, infrastructure—each depending on time that agrees with itself everywhere, all at once.

The discrepancy appeared first as an accounting problem.

If Earth rotated a little slower than yesterday, then civil time would eventually drift away from astronomical time. Midnight would no longer mean what it once did. The sky would slip relative to the clock.

So adjustments appeared.

Leap seconds were added. Single, awkward moments inserted into official timekeeping to keep the human calendar aligned with the planet's rotation. Clocks paused. Systems were patched. Engineers wrote workarounds. Most people never noticed, except when something broke.

And things did break.

Software stumbled. Networks desynchronized. Systems designed for smooth continuity balked at the idea that time itself might hesitate. Each correction carried a cost.

The infrastructure had been built on the assumption that time was uniform and obedient.

Arguments followed.

Were leap seconds worth the trouble? Was it better to let clocks drift and correct later? Should civil time remain tethered to Earth's rotation, or should it become an abstract standard, free of celestial inconvenience?

None of these questions were philosophical. They were operational.

Eventually, decisions were made to stop adding leap seconds—not because the Moon's pull had weakened, not because Earth had stabilized, but because managing the coupling had become disruptive. The cost of honoring the system exceeded the tolerance of the systems built on top of it.

Time itself began to require judgment.

The world it described had become too interconnected to ignore small mismatches. Earth's rotation was influenced by oceans, atmosphere, core dynamics, and the Moon's persistent tug. No single cause could be isolated cleanly enough to assign blame or simplicity.

The system was coupled.

Separation was no longer possible without consequence.
Intervention became a matter of choice rather than correction.

To the public, this arrived stripped of context.

A headline about leap seconds being eliminated.
A joke about clocks being weird.
A vague explanation that "time is complicated."

It was treated as trivia—another technical oddity handled by experts somewhere else. Something to laugh about, then forget.

The clocks kept needing human judgment.

Few people asked why.

Fewer still connected that need to a slow exchange of energy between Earth and Moon, measured quietly for decades, accumulating its effects one millisecond at a time.

Time drifted.
The Moon drifted.
The system adjusted.

And because nothing catastrophic happened, the adjustment itself became invisible—another sign absorbed without meaning, filed away as an inconvenience rather than a consequence of a coupled world slowly revealing itself.

∝ ∝ ∝

For sixty years, the signs were present.

They didn't arrive as revelations. They arrived as measurements—quiet, disciplined, repeatable. Numbers taken seriously enough to be published, checked carefully enough to be trusted, and refined patiently as instruments improved. Each decade sharpened the picture. Each generation inherited a cleaner version of the same result.

Measured.
Published.
Repeated.
Refined.

There was no missing data, no gap in observation that could later be blamed for misunderstanding. The record was continuous. The signal never went away. It simply never changed character enough to demand attention.

As the results escaped their original contexts, they leaked into public awareness the way technical facts usually do—incrementally and without explanation, through simplification. A number here. A sentence there. A diagram stripped of its caveats and displayed as reassurance.

They were not presented as warnings because they didn't behave like warnings. There was no event to point to, no image that demanded replay, no deadline embedded in the phenomenon itself. Scale did the rest of the work. When change stretches across centuries, urgency dissolves.

Familiarity completed the process. Repetition softened edges that had once felt sharp, and what was once notable became expected. Expectation neutralized urgency more effectively than doubt ever could.

Once a fact is repeated often enough without consequence, it becomes harmless. It loses the ability to startle. People learn where to store it mentally—alongside trivia, curiosities, and background knowledge that exists to be admired rather than acted upon.

Nothing here was secret.

The measurements were public. The papers were accessible. The explanations were taught in classrooms and narrated in documentaries. There was no moment where the truth had to be smuggled out or dragged into the open.

Nothing here was denied.

No one stood up and said the Moon wasn't moving, or that time wasn't drifting, or that the systems were not coupled. Each fact was acknowledged on its own terms, then allowed to settle quietly back into the background.

But meaning requires assembly.

It does not emerge automatically from accuracy. It requires connection, framing, and a reason to believe that disparate facts belong to the same story. Without that, even the clearest truths remain inert.

By the time the public encountered these fragments, they arrived without the scaffolding that gave them weight. Each piece appeared isolated, divorced from the others, stripped of the relationships that made them consequential. A drifting Moon here. A leap second there. Precision science as spectacle. Timekeeping as inconvenience.

No synthesis was offered.

No one explained how these facts spoke to one another, or why their convergence mattered more than their individual harmlessness. There was no moment that forced the pieces to cohere into a single picture demanding response.

So they became background noise.

Not false or ignored.

Simply unassembled.

That was the translation gap.

Not a failure of truth—but a failure of reception.

∝ ∝ ∝

What Changes Over The Next Hundred Years (Realistic Projections)

Over the next hundred years, nothing dramatic happens.

That is the hardest part to explain.

Because nothing happening feels like reassurance.

The Moon will be measurably farther away—not in a way bodies can see, not in a way tides suddenly betray, but in a way instruments, and people, cannot ignore anymore. At roughly a few centimeters per year, the accumulation is small enough to sound dismissible and large enough to be undeniable once you insist on counting.

In a century, the distance grows by a few meters.

Not kilometers. Not something you photograph. Not something anyone points at the sky and notices. The Moon looks the same. It rises and sets as it always has. No calendar needs revision. No night feels different.

And yet the numbers change.

Change does not require visibility.

Laser returns arrive a little later. Orbital parameters require slightly different corrections. Models built on yesterday's constants need gentle updating, the way engineers quietly update tolerances when materials age. Nothing breaks. Nothing announces itself.

Humans will not *see* this change.

Civilizations will measure it.

Measurement is a form of inheritance.

That difference matters.

A few meters in a century sounds trivial until you remember how many centuries a civilization assumes when it builds calendars, navigation systems, treaties, and expectations into its infrastructure. Assumptions about stability are rarely written with expiration dates. They are inherited, reused, and trusted precisely because nothing obvious ever goes wrong.

The Moon's retreat does not threaten lifetimes.

It tests continuity.

Continuity is where institutions fail slowly.

The change is slow enough to invite dismissal, fast enough to accumulate across institutional memory. Each generation inherits a slightly different sky, quantified so precisely that denying the shift would require effort—but accepting its implications would require something harder.

Synthesis.

Without it, accumulation remains inert.

This is how the next hundred years unfold: not as revelation, but as quiet divergence between what is seen and what is known. The Moon will be farther. The data will say so. And the difference between those two statements will remain as the space where meaning struggles to form.

Over the next hundred years, the Moon's pull weakens a little.

Not enough to redraw coastlines.

Not enough to change tides in a way anyone standing on a beach could point to and name. High tide still comes. Low tide still follows. Harbors don't empty. Cities don't flood or dry.

The shorelines don't notice.

Life continues to meet the water where it always has. The boundary holds, even as the balance behind it shifts.

Stability can conceal redistribution.

But the force that shapes them—slow, rhythmic, gravitational—does not disappear cleanly. It thins.

Tidal forcing depends strongly on distance, and distance, even when it grows by only a few meters, compounds its effect over time. The change is subtle enough to vanish inside weather variability, storm cycles, seasonal shifts. No single tide is wrong. No chart spikes. No alarm sounds.

But in long records, the pattern begins to soften.

It shows up first where timing matters more than magnitude.

Coastal ecosystems that evolved to synchronize with lunar cycles begin to drift out of alignment—not dramatically, not catastrophically, but persistently. Breeding events keyed to tidal extremes arrive slightly off cadence, their timing no longer matching historical patterns with the same reliability they had. Larval dispersal windows widen or narrow by margins too small to notice in a single season and too consistent to ignore over decades.

Fish don't fail to spawn.

Crabs don't vanish.

Coral does not collapse overnight.

But populations adjust.

Adjustment is not neutrality.

Ecologists notice before urban planners do because they work with long baselines and cumulative patterns. Their data does not ask whether tides still exist. It asks whether timing still optimizes survival.

The answer shifts slowly.

A nesting season stretches.

A migration window drifts.

Recruitment rates wobble within "normal" bounds—but normal itself is inching sideways.

No press release accompanies this. No headline announces *Moon Weakens Tides*. The attribution problem is too complex, the signal too distributed. Oceans respond not just to the Moon, but to winds, currents, temperature, basin geometry. No single cause can be isolated cleanly enough to claim ownership of the change.

That ambiguity protects the narrative.

From a human perspective, nothing has altered. Storms still do more damage in an afternoon than this effect does in a decade. Infrastructure remains dominant. Engineering still overwhelms subtle shifts in gravity.

Cities feel nothing.

But biological systems are less buffered. They integrate small forces relentlessly. They carry memory forward in generations, not reports. They respond to drift not by noticing it, but by reshaping themselves around it.

By the time the change becomes legible to humans, it will not appear as a tidal problem at all.

It will appear as ecology behaving differently.

And ecology, when it changes slowly, rarely points to the sky.

The shorelines don't notice.

The breeding cycles do.

∝ ∝ ∝

Over the next hundred years, time doesn't settle down.

If anything, it becomes harder to pretend that it ever was simple.

Earth's rotation continues to evolve, slowly and unevenly. The Moon's long-term tidal pull remains part of the story, exchanging energy the way it always has, lengthening days by imperceptible margins. But it is no longer the only voice that matters. Oceans shift mass. Ice melts and redistributes weight. Atmosphere and core trade momentum in ways that are measurable but not cleanly predictable.

The result is noise riding on drift.

Some years Earth spins a little faster than expected, registering small deviations that disappear in noise; other years a little slower than models predict, with no cause standing out. The long-term trend persists beneath the variability, but the surface remains uneven enough to frustrate anyone trying to impose order on it. Atomic clocks continue counting with perfect indifference. Earth does not.

For systems that demand synchronization, the mismatch grows irritating.

Small discrepancies accumulate at interfaces.
Tolerance shrinks where precision is assumed.
What was once negligible becomes operational.

Civil time drifts against astronomical time. Corrections become awkward. Engineers debate whether to preserve alignment with the sky or to sever the link entirely and let clocks become purely abstract. Every option solves one problem and creates another.

The arguments are technical, but the consequences are not.

Leap seconds linger as a symbol long after their practical usefulness is questioned. They are too small to matter to daily life and too disruptive to ignore in global systems. Networks fail when they are mishandled. Markets pause. Navigation stumbles. No one loses a day—but everyone loses patience.

Committees form.

Reports circulate.

Some argue that civil time should remain tethered to Earth's rotation, that the sky should still mean something when the clock says midnight. Others argue that modern infrastructure cannot afford to negotiate with a planet that refuses to keep time cleanly.

Neither side is wrong.

That is the problem.

Because correctness offers no refuge.

Because the physics does not offer a resolution. The Moon continues pulling. Earth continues responding unevenly. Climate-driven mass redistribution adds variability on top of old cycles. The system remains coupled, dynamic, and inconvenient.

So the question shifts.

Not *what time is it?*

But *which definition are we willing to live with?*

Time becomes something that must be decided.

Not discovered.

Not measured alone.

Administered.

Every choice embeds values—about tradition, precision, tolerance for drift, and who bears the cost when synchronization fails. These are not scientific disagreements. They are governance problems masquerading as technical ones.

Most people never notice the debate directly. They only notice when updates are announced, when standards quietly change, when software patches roll out to handle yet another exception. Time remains familiar enough to trust and strange enough to require constant explanation.

It keeps working.

But it no longer feels neutral.

The next century does not bring temporal chaos. It brings something more subtle and more persistent: the normalization of argument. Timekeeping becomes a recurring agenda item, a policy issue revisited every few years, a reminder that even the most basic assumptions of daily life now require ongoing negotiation.

Time becomes an administrative argument. Administration replaces inevitability with choice.

And once time itself needs committees, it becomes harder to pretend that the systems built on it are standing on anything immutable at all.

∝ ∝ ∝

Over the next hundred years, the most important thing is what **doesn't** happen.

Nothing dramatic arrives.

The Moon does not disappear. The sky does not tilt. Seasons don't slip into unfamiliar patterns. There is no sudden instability in Earth's axis, no cinematic unraveling that forces people to look up and reconsider what they thought was permanent.

The fears people reach for instinctively—the ones that belong to documentaries and speculative conversations—remain far beyond a human century. The Moon continues to do what it has always done on timescales that matter to lived experience. Climate does not suddenly lose its rhythm because of orbital distance alone. The planet keeps its posture.

Anyone waiting for a moment that announces *now* will keep waiting.

And that absence matters.

Absence denies closure.

Because without a catastrophe, there is nothing to rally around. No single failure collapses uncertainty into urgency. No event forces agreement. Systems continue to function well enough to justify confidence, at least locally, at least for now.

The tension stays where it has always lived.

Inside institutions that quietly revise assumptions instead of questioning them.

Inside risk models that grow more complex without ever becoming more decisive.

Inside language that remains careful, accurate, and increasingly disconnected from synthesis.

Public explanations stay calm because nothing visible contradicts them. Experts continue to hedge because the data still allows it. No one is lying. No one is wrong enough to be discredited.

What changes instead is subtle and difficult to name.

Confidence does not collapse.

It thins.

Not confidence in physics.

Not confidence in measurement.

Confidence in permanence.

The sense that the systems inherited from the past were built on constants that didn't require revisiting begins to fade. Systems demanded more maintenance than anyone expected. Corrections accumulate. Exceptions multiply. Review cycles return sooner than planned.

The future starts to feel less like a destination and more like a series of extensions—granted, provisional, always subject to reassessment.

Nothing ends.

The catastrophe never arrives.

What leaves, quietly, is the belief that stability was automatic—that it could be inherited indefinitely without attention, without judgment, without choice.

And once that belief is gone, the world does not feel endangered.
It feels provisional.
Which is far harder to live with.

PHASE 5 — POLITICAL FRACTURE: TRUTH VS "INCOMPLETENESS"

The fracture didn't follow familiar lines.

It didn't sort itself into parties, ideologies, or slogans that could be mapped cleanly onto a spectrum. There were no banners for it, no chants, no simple way to explain who stood where without sounding evasive or abstract.

What split instead was something quieter.

A disagreement about risk.

It wasn't a dispute about facts, but about sequence. About whether knowing must come before preparedness, or only after it. About who bears the cost of disclosure when consequences unfold slowly. About whether restraint protects the public—or protects institutions from having to choose. And about how long caution can pass for care before it becomes something else.

On one side were those who believed that truth, once sufficiently established, carried an obligation to be spoken—even if it arrived incomplete, conditional, and resistant to reassurance. Withholding action felt like a decision made on someone else's behalf. They argued that uncertainty wasn't a flaw to be hidden, but a condition to be shared.

On the other side were those who believed stability itself was a public good—fragile, cumulative, and easily damaged by premature disclosure. They didn't dispute the data. They didn't deny the implications. They argued that releasing truths without resolution invited reaction without remedy, fear without agency, and pressure without options.

Both positions were defensible.

Defensible is not resolvable.

That was the problem.

Because legitimacy existed on both sides.

In closed rooms and policy briefings, the same phrases began to surface again. *Incomplete. Preliminary. Context-dependent.* Words that were technically correct and politically powerful, well matched to careful presentation and selective emphasis. Each carried a different weight depending on who heard them. Each could be used to justify restraint or to accuse delay.

The argument was never about whether the science was real.

It was about custodianship.

It was about when reality became actionable—and who got to decide.

Those who favored disclosure spoke in the language of responsibility. They worried about precedent, about what it meant to normalize silence once it became convenient. They feared that delaying truth trained institutions to confuse caution with care.

Those who favored restraint spoke in the language of harm reduction. They worried about misinterpretation, about markets and movements reacting faster than understanding could keep up. They feared that releasing incomplete truths would fracture trust faster than any slow, measured drift ever could.

Neither side saw itself as political.

Both sides accused the other of being reckless.

And because the disagreement wasn't about goals but about timing, it resisted compromise. You couldn't split the difference between speaking and waiting. You could only choose which risk felt more tolerable.

The fracture spread unevenly.

Scientists found themselves aligned not by discipline, but by temperament. Administrators by experience. Advisors by what they had seen break before. Even language began to diverge—some leaning into precision and caveat, others into synthesis and framing.

What emerged wasn't polarization, but misalignment.

Misalignment is governance under strain.

A system built to absorb uncertainty now had to decide whether uncertainty belonged inside its walls or outside them.

And for the first time, the question was no longer whether the truth could be measured.

It was whether it could be released without destabilizing the very structures tasked with responding to it.

That was the political fracture.

A fracture in authority, not evidence.

Not truth versus falsehood.

Truth versus incompleteness—and the fear of what happens when the public is asked to hold both at once.

∝ ∝ ∝

Two Camps Emerge (Quietly, At First)

Camp A formed quietly.

Quiet is how systems divide.

No one announced it. No memo circulated. The phrase *Disclosure / Continuity Realists* came later, after it was already obvious that the same people kept ending up on the same side of the table.

They were the ones who stayed after meetings.

Not to argue louder—but to keep arguing at all.

One evening, long after the official briefing had ended, a half-dozen people lingered in a conference room that smelled faintly of burnt coffee and dry-erase markers. Screens had gone dark. The agenda had been cleared. The discussion, finally, could breathe.

"We keep acting like disclosure is escalation," said Dr. Lena Kovács, folding her arms. "It isn't."

Across from her, Jonah Price—barely thirty, already buried in policy memos—shook his head. "No. It's *decompression.* You let air out slowly so the system doesn't rupture later."

Someone laughed softly at that. Not mockery. Recognition.

"Say that in the briefing," an older man said. Ravi Menon, agricultural systems forecaster, hair gone white long before retirement. "See how it lands."

Lena didn't smile. "I'm serious. We're treating 'nothing dramatic is happening' like a reason to stay silent. That's backwards."

A woman near the window—climate-adjacent, water systems, used to being ignored until it was too late—leaned forward. "If something dramatic *were* happening, we'd all agree disclosure was necessary. The only difference now is that we still have time."

Jonah nodded. "That's the whole point. Time is the asset. Silence spends it."

Ravi tapped the table once, deliberately. "In my world, slow risk is the worst kind. Crops don't fail all at once. They degrade. Yield slips. Margins thin. By the time politicians notice, adaptation windows are gone."

"So you want to tell the public what?" someone asked from the doorway. "That the Moon is moving?"

"No," Lena said sharply. "We want to tell them the *system* is coupled. That slow change is real. That nothing is breaking—but nothing is frozen either."

Jonah added, "We're not asking for panic. We're asking for expectation management."

"That's not how it'll be received," a voice said from the back. Cautious. Experienced. "People hear uncertainty as danger."

"They hear it as danger when it arrives suddenly," Lena replied. "When it's been hidden and then forced into the open by consequences."

Ravi leaned back. "I've seen this before. With groundwater. With soil salinity. With heat thresholds. We always say, 'Not yet. Not mature enough.' Then one day we say, 'Why didn't anyone prepare?'"

Silence followed—not disagreement, but discomfort.

Discomfort is where policy begins.

Jonah broke it. "We disclose *because* nothing dramatic is happening yet. That's the argument. That's the line."

Someone wrote it down.

Writing makes a position real.

The camp wasn't unified by temperament. Some were cautious speakers. Some were blunt. Some had scars from earlier fights they'd lost. What bound them was a shared refusal to confuse *absence of crisis* with *absence of obligation*.

They didn't believe disclosure meant certainty.

They believed it meant honesty about trajectory.

They argued that societies deserved time—real time, not emergency time—to adapt their assumptions. That trust wasn't preserved by silence, but by context delivered early enough to become familiar rather than shocking.

Silence, in their view, didn't prevent harm.

It deferred it.

And deferral, they knew, always collected interest.

Quietly at first, the same names kept appearing in the same threads, the same side conversations, the same unscheduled follow-ups. Not a rebellion. Not a movement.

A camp.

A camp is alignment made visible.

∝ ∝ ∝

A camp.

That word made some of them flinch when it was first used out loud.

"We're not factions," Lena said, the first time she heard it. "We're just… agreeing."

Jonah shrugged. "Agreement becomes a camp the moment someone notices it."

They noticed it quickly.

Emails that once went to broad lists began to loop the same names. Invitations to "informal check-ins" landed with high predictability.

Draft language circulated privately before it ever reached official channels, annotated not with edits but with questions: *How would this land? What does this assume people already know? What breaks if we say this now instead of later?*

The conversations lengthened.

They stopped being about the Moon specifically and started being about precedent.

"If we don't contextualize this," Ravi said one afternoon, staring at a projection no one was really looking at anymore, "we're teaching ourselves that slow truths don't count."

"Slow truths always count," Jonah said. "They're just inconvenient."

"That's not the same thing."

"It is to institutions."

Lena leaned forward. "What we're really arguing is whether incompleteness disqualifies honesty."

A pause followed that. The kind that meant the question had landed.

Some questions reorganize rooms.

"Incompleteness disqualifies *authority*," someone said finally. A transportation systems analyst who had joined late, quiet until now. "Not honesty."

"Authority over what?" Lena asked.

"Over reaction," he replied. "Over what happens next."

Jonah nodded slowly. "That's the fear, isn't it? That if we say it out loud, we lose control of the timeline."

"And if we don't," Ravi said, "we lose control of the outcome."

They began to trade examples, not rhetorically, but almost therapeutically.

Groundwater basins where drawdown was known for decades before wells ran dry. Bridge ratings quietly downgraded until a collapse forced a reckoning. Climate thresholds discussed cautiously until they crossed.

"We always say the same thing," Lena said. "*The data wasn't mature enough yet.*"

"And sometimes it wasn't," someone objected.

"Yes," she said. "But sometimes it was mature enough to explain *direction*, and we waited for precision we didn't actually need."

Jonah pulled up an old memo on his tablet. "This was written twelve years ago. Same structure. Same language. Same deferral."

Ravi smiled without humor. "We're very good at recycling caution."

What surprised them wasn't the resistance—they had expected that—but how often the resistance came from people who privately agreed with them.

A deputy director took Lena aside after a briefing and spoke in a low voice. "You're not wrong."

Correctness is not protection.

"Then why—"

"Because if this goes sideways," he said, "it won't be the data that's blamed. It'll be the decision to speak."

"That's always true," Lena said.

"Yes," he replied. "But it's not always survivable."

The camp absorbed that reality without surrendering to it.

They were not reckless. They didn't advocate dramatic announcements or simplified narratives. In fact, they spent more time arguing among themselves about phrasing than they did with anyone outside.

"How do you say 'this matters' without implying 'this is urgent'?" Jonah asked one night.

"You don't," Ravi said. "You say 'this is cumulative.' People understand accumulation."

"Do they?" someone asked.

"They understand interest," Ravi replied. "They understand maintenance. They understand rot. They just don't like admitting it applies to big things."

They rehearsed sentences the way trial lawyers rehearsed openings.

This is not a crisis.

This is not new.

This is not a prediction.

None of it felt sufficient.

Because the real argument wasn't semantic.

Language was only the battlefield.

It was philosophical.

One evening, the conversation drifted—as it often did—into time horizons.

"Most of our systems are built on the assumption that someone later will fix it," Lena said. "Disclosure is how you tell people that *later is now part of the equation.*"

Jonah nodded. "We're not asking for action. We're asking for orientation."

"And that," Ravi said, "is exactly what scares them."

They began to see the pattern clearly then. The resistance wasn't about fear of panic. It was about fear of responsibility spreading too widely.

Once the public was oriented—once slow change was acknowledged as real, ongoing, and relevant—expectations shifted. Questions followed. Accountability diluted upward and outward.

Expectation is a form of leverage.

Institutions preferred responsibility to remain centralized, manageable, paced.

Disclosure threatened that.

Still, the camp grew.

Not loudly. Not officially. But steadily.

A hydrologist forwarded a paper "for context."

An infrastructure planner added a quiet footnote to a draft.

A communications officer asked, off the record, "If we *were* to explain this, how would you do it?"

They answered patiently.

Again and again.

Always the same core belief, expressed in different ways:

Stability comes from shared understanding, however unfinished.

And delay, however justified, accumulates consequences of its own.

Silence does not preserve stability.
It postpones adaptation.
And postponement is itself a choice.

By the time someone finally said it plainly—during a tense interagency meeting, microphones on, minutes being taken—it no longer sounded radical.

"We disclose," Lena said, voice steady, "because nothing dramatic is happening yet."

A few people shifted uncomfortably. A few nodded. One person wrote it down.

The camp didn't cheer.

They had learned, by then, that agreement didn't mean victory.

It meant the argument had become unavoidable.

Unavoidable arguments produce structure.

And once an argument reaches that stage, the system does what it always does next.

It forms the opposing camp.

∝ ∝ ∝

Camp B didn't think of itself as opposition.

Opposition is a label applied later.

If anything, its members believed they were the adults in the room.

They had been in those rooms longer than most. They knew how easily systems could tip—not in theory, but in markets, in alliances, in supply chains that behaved well only as long as confidence held. Their instinct wasn't denial. It was containment.

The argument surfaced most clearly in rooms where the walls were thicker and the schedules tighter.

A senior treasury advisor leaned back in his chair, fingers steepled, listening without interrupting as the presentation wound down. When it finally did, he waited another beat before speaking.

"Let me ask a simple question," he said. "What does the public *do* with this?"

No one answered immediately.

"I don't mean what do they *understand*," he continued. "I mean what action follows disclosure?"

Lena started to respond, but he held up a hand.

"If the answer is 'nothing,'" he said, "then what we are doing is transferring uncertainty from institutions designed to hold it to a public that isn't."

Across the table, a defense-linked advisor nodded. "Uncertainty is not an emergency. Treating it like one creates emergencies."

Jonah frowned. "No one is calling it an emergency."

"But the public hears thresholds," the advisor replied. "They hear drift. They hear loss of control. You don't get to decide how that narrative evolves once it's out."

A market analyst cleared her throat. "Markets don't price centuries. They price quarters. You release a truth with no actionable response, you don't get patience. You get volatility."

"And volatility," the treasury advisor added, "isn't philosophical. It's mechanical."

They were not alarmists. They didn't raise their voices. They spoke with the confidence of people who had seen what happened when abstract risks escaped into systems built on expectation.

"We are not disputing the data," a senior politician said quietly. "We are disputing timing."

Ravi leaned forward. "Timing *is* the decision."

"Yes," the politician said. "And it's one we're paid to make."

That was the difference.

Authority was the stake.

Camp B believed responsibility meant deciding *for* the system when exposure created more harm than benefit. They didn't deny slow risk. They contextualized it as background—real, measurable, but not yet actionable.

"Actionability matters," the defense advisor said. "You don't brief the public on every long-term strategic imbalance unless there's a lever to pull."

"And what if the lever is preparation?" Lena asked.

The advisor shook his head. "Preparation without a defined objective is just anxiety with a budget."

They worried about precedent as much as Camp A did—but in the opposite direction.

"If we disclose this," the treasury advisor said, "what do we say next year when the numbers haven't changed meaningfully?"

Jonah answered, "We say the same thing."

"And the year after?"

"The same."

"And after ten years of 'the same,'" the advisor replied, "what credibility do we have left when something *does* change quickly?"

Silence followed.

Camp B had lived through crises where early warnings had been technically correct and politically catastrophic. Where incomplete truths had outrun institutions' ability to respond. Where trust had collapsed because officials spoke too soon.

They remembered runs on banks triggered by statements meant to reassure. Markets that reacted to tone rather than content. Allies who interpreted disclosure as weakness or instability.

"Institutions exist to absorb uncertainty," the politician said. "That's their function. When we externalize it prematurely, we abdicate that role."

Lena replied carefully. "And when does absorption become hoarding?"

"When release creates remedies," he said without hesitation. "That's the line."

They believed deeply in remedies.

Not symbolic gestures. Not awareness. Concrete mechanisms—policy levers, economic buffers, international coordination frameworks. Without those, disclosure felt irresponsible, even cruel.

"People don't need truths they can't act on," the market analyst said. "They need stability."

Ravi countered, "They need time."

"And time," she replied, "is not neutral in markets."

Camp B didn't see silence as deception.

They saw it as stewardship.

Stewardship can resemble control.

To them, saying *not yet* wasn't avoidance—it was protection. Of trust. Of continuity. Of systems that depended on the assumption that tomorrow would resemble today closely enough to plan.

They didn't believe the public was fragile.

They believed the system was.

And systems, once destabilized, didn't wait politely for clarification.

In private, some of them admitted discomfort.

"This isn't forever," the politician said once, late, when only a few remained. "It's sequencing."

"But sequencing toward what?" Lena asked.

He paused. "Toward something we can do."

No one accused anyone of lying.

No one disputed the science.

The fracture was never about truth versus falsehood.

It was about *who* should hold uncertainty, *for how long*, and *at what cost.*

And because both camps believed they were protecting continuity, neither could easily yield.

The disagreement hardened—not into hostility, but into something more durable.

A structural divide.

Structure outlasts personalities.

One side believed time was the solution.

The other believed time was the problem.

And the system, caught between them, did what it always did when faced with incompatible philosophies.

It waited.

Waiting is also a policy.

∝ ∝ ∝

The Phrase That Stabilizes The Era

The phrase appeared quietly.

Not as doctrine.

Not as policy.

Just as language—repeated often enough to feel inevitable.

It surfaced first in draft reports, slipped into executive summaries as a qualifier that seemed responsible rather than evasive. A sentence would end cleanly, then soften at the edge.

The information is technically accurate, but incomplete.

Accuracy became permission to postpone.

No one objected to it. In fact, it soothed people. It sounded honest. It sounded careful. It sounded like science behaving properly.

In one briefing, a press officer read it aloud, then looked up.

"We're comfortable with that wording, yes?"

A few heads nodded.

"It's true," someone said. "We're not overstating."

"And we're not hiding anything," another added. "The data's there."

No one mentioned that the phrase did more than describe the state of knowledge.

It managed responsibility.

Management is not resolution.

Soon it became reflex.

A reporter asked whether the findings suggested long-term structural change.

"Yes," came the answer, "but the information is technically accurate and incomplete."

Another asked whether there were implications beyond current models.

"There may be," the spokesperson said carefully. "But again—technically accurate, incomplete."

The words were always delivered calmly, with practiced neutrality. They were not defensive. They didn't sound like denial. They sounded like professionalism.

And they worked.

Because they ended conversation without ending knowledge.

"Incomplete" didn't invite follow-up. It closed the loop with the appearance of process intact, signaling that the matter remained under review with no escalation or attention required. It suggested that questions had been asked, that answers existed, and that the appropriate next step was patience—just not yet.

Inside institutions, the phrase spread faster than any memo.

Analysts used it to protect nuance.

Administrators used it to protect timelines.

Politicians used it to protect stability.

Everyone could agree to it, because it asked nothing of them in return.

In one meeting, Jonah leaned back and said quietly, "We should be careful with that word."

"Why?" someone asked.

"Because it doesn't age," he said. "It can mean six months. It can mean fifty years."

"That's the point," the reply came. "It keeps us flexible."

Lena frowned. "It keeps us *unaccountable.*"

No one answered that directly.

Because "incomplete" had become more than a descriptor.

It was a shield.

A shield against consequence.

It allowed reports to remain technically correct while avoiding synthesis. It allowed briefings to proceed without escalation. It allowed nothing to break.

When questioned later—by auditors, by journalists, by internal reviews—the same sentence could be invoked retroactively.

We said it was incomplete.
We never claimed finality.
We were transparent.

And all of that was true.

What the phrase quietly prevented was closure.

Closure would have required ownership.

As long as information remained incomplete, no one had to decide what completeness would look like—or who would be responsible for declaring it achieved. The future could remain provisional. The present could remain stable.

"Incomplete" absorbed tension the way a shock absorber absorbs force. It dulled impact. It preserved motion.

Over time, people stopped noticing it at all.

It became the background hum of official language—the soft qualifier that followed every uncomfortable truth and ensured it arrived without consequence.

The information was accurate.

That mattered.

But incomplete meant unfinished, and unfinished meant later, and later meant not now.

And as long as *not now* could be justified in a single word, the era had its defining phrase.

Nothing broke.

Nothing had to.

"Incomplete" made sure of that.

One word became infrastructure.

∝ ∝ ∝

The word began to travel on its own.

It appeared in places no one could trace back to a decision. Slide decks cloned from past meetings and revised each time. Margin notes in drafts that never reached publication. Someone would add it reflexively, the way people add dates or units—part of the grammar of caution.

Technically accurate, but incomplete.

Soon it was no longer necessary to say the whole phrase. *Incomplete* alone carried the weight.

A junior analyst hesitated during a briefing. "Should we mention the long-term coupling here?"

A supervisor glanced at the page. "Flag it as incomplete."

That was enough.

In press rooms, the effect was immediate and predictable. Reporters learned the boundary quickly. Incomplete meant there would be no follow-up sentence worth chasing. It meant the answer wouldn't harden into a headline. It meant the story would fail the test of urgency.

"So you're saying this could matter?" one journalist asked, leaning forward.

"It could," the spokesperson replied evenly. "But the information is incomplete."

"On what timescale?"

"Long."

"How long?"

"Long enough that no immediate action is indicated."

The reporter wrote something down, then stopped. "So… nothing to do."

"Not at this time."

The article never ran.

Inside the institutions, the word did something subtler. It reorganized responsibility.

As long as information remained incomplete, accountability remained diffuse. No office owned the implications. No department had to propose remedies. The work could remain analytical rather than operational.

One committee chair said it plainly, once, after most of the room had emptied.

"Incomplete buys us time."

"And then?" Lena asked.

"And then we reassess."

"On what basis?"

He hesitated. "On whether it's still incomplete."

The circularity went unremarked.

Because *incomplete* didn't demand a threshold. It didn't require a metric for sufficiency. It didn't specify what would change its status. It simply held the present open, indefinitely.

The phrase crept into internal reviews.

No material deviation observed.

Findings remain consistent with prior models.

Implications are still incomplete.

Each sentence was defensible. Each was accurate. Together, they formed a kind of stasis—an equilibrium of postponement.

Even those who disliked the word found themselves using it.

Lena caught herself once, halfway through a sentence. She stopped, frowned, then finished it anyway.

Later, Jonah noticed.

"You used it," he said quietly.

"I know," she replied. "That's what scares me."

Because once a word becomes infrastructure, it stops feeling like a choice.

"Incomplete" didn't silence anyone outright. It didn't forbid discussion. It simply slowed everything down enough that urgency evaporated on contact. Arguments softened. Edges rounded off. The future remained abstract.

Years later, when auditors reviewed old reports, they found no lies.

The information had been accurate.
The caveats had been explicit.
The limitations had been acknowledged.

Everything had been done correctly.

What no document recorded was how long *incomplete* had been allowed to mean *later*, and how much adaptation that later had quietly displaced.

By the time anyone asked when the information would be considered complete, the question itself sounded naïve.

Complete relative to what?
Complete for whom?
Complete enough to do what, exactly?

"Incomplete" had done its job.

Nothing had broken.
Nothing had forced a reckoning.
Nothing had required the system to decide whether knowledge carried obligations beyond description.

The era didn't announce itself.

It simply learned how to say *not yet* in a way that sounded responsible—and stopped noticing how often it did.

∝ ∝ ∝

The phrase began shaping meetings before anyone noticed.

Agendas shortened. Conclusions softened. Action items transformed into monitoring plans. Every proposal that drifted too close to implication was nudged back toward analysis.

"Let's not outrun the data," someone would say.

"What data?" Jonah asked once.

"The complete data."

"And when will that be?"

A pause. Then, lightly: "When it's ready."

No one laughed.

"Incomplete" became a way to end discussions without appearing to shut them down. It allowed disagreement to dissolve into patience. It made urgency feel impolite.

A defense advisor used it to steady a room that had grown tense.

"We all agree the signal is real," she said. "We also agree it's incomplete. Until that changes, our posture shouldn't."

"And what is our posture?" Lena asked.

"Stable," came the answer. "Predictable."

Later, in the hallway, Jonah muttered, "Predictable for whom?"

The phrase started appearing in emails marked *For Awareness Only.*

Findings remain technically accurate but incomplete. No change to operational stance.

That sentence did more work than pages of analysis ever had.

It traveled upward easily. Senior offices liked it because it preserved optionality. It could be cited to justify restraint today and foresight tomorrow. It committed no one to anything except continued observation.

It traveled downward just as smoothly. Analysts learned which conclusions would be welcomed and which would be softened. Graduate researchers learned where synthesis stopped and speculation began. Everyone learned the boundary without it ever being drawn explicitly.

A communications officer said it plainly during a training session.

"If you use the word 'incomplete,' you're protected."

"From what?" someone asked.

"From follow-up."

The room went quiet.

Over time, the phrase began to come into view preemptively—inserted before anyone asked for it. Reports anticipated discomfort and cushioned it in advance. The qualifier arrived so early in sentences that readers barely registered the content it was qualifying.

Technically accurate.

Incomplete.

The order mattered.

Accuracy reassured. Incompleteness excused.

In one closed-door review, a senator flipped through a briefing book, stopped, and tapped the page.

"So we're saying this matters," he said, "but not enough to do anything."

"We're saying it matters," the advisor replied carefully, "but not enough to change policy."

The senator nodded. "Good. Because change is expensive."

No one argued.

What made the phrase so powerful was that it didn't sound like refusal. It sounded like care. Like professionalism. Like restraint earned through experience.

Even those who opposed its use found themselves trapped by its logic.

"If we stop calling it incomplete," Lena said once, "what are we calling it instead?"

Ravi answered quietly. "Ongoing."

"That's worse," someone said. "That sounds endless."

"Because it is," Ravi replied.

The room didn't accept that easily.

"Incomplete" suggested a finish line. It implied that one day, someone would stand up and say *now it's complete.* It allowed everyone to imagine closure without defining it.

But no one ever volunteered to name that day.

Years passed. New administrations came and went. Staff rotated. The phrase remained.

When questions resurfaced—inevitably—they were met with familiar language, recycled seamlessly into new contexts.

Nothing had changed enough.

The picture was still forming.

The information remained incomplete.

By then, it no longer felt like a decision.

It felt like gravity.

And like gravity, it acted constantly, invisibly, shaping motion without ever announcing itself.

The system didn't lie.

It did something more subtle.

It learned how to tell the truth in a way that required nothing in return.

∝ ∝ ∝

Why Disclosure Is Resisted

Disclosure wasn't resisted because it was false.

It was resisted because of what followed once people *understood.*

Not the headline version. Not the simplified takeaway. The second-order understanding—the one that settled in after the initial explanations wore off and the implications began to connect.

If the Moon mattered to Earth's long-term stability, not symbolically but mechanically—if that stabilizer was slowly weakening, not catastrophically but persistently—
if there was no lever to pull, no engineering solution waiting just beyond the horizon—
and no timeline that fit inside an election cycle—

then something fundamental shifted.

Not emotionally.

Structurally.

The first reaction wasn't panic. It was recalculation.

Investors began asking questions that had nothing to do with returns next quarter and everything to do with horizons they had never needed to articulate before. What did *long-term* mean, exactly, if the systems underwriting stability were themselves in motion? At what point did foresight turn into speculation?

Projects didn't stop.

They hesitated.

Infrastructure plans added contingencies. Timeframes stretched. Assumptions that had once been implicit now demanded justification. Someone would ask, quietly, in a meeting meant to be routine, "Are we assuming continuity here—or betting on it?"

No one liked answering that.

Political language began to thin.

Promises were still made, but they lost profundity. They sounded complete on the surface and unfilled underneath. Terms like *sustainable* and *permanent* required footnotes. Policy platforms grew careful, then vague, then defensive.

"What does sustainable mean," a staffer asked during a drafting session, "if stability itself is conditional?"

The room went still.

Economic models, especially the ones built on discounting the future into near irrelevance, began to feel exposed. Not broken—just provisional. Their logic still held, but their confidence didn't. They worked as long as tomorrow resembled today closely enough.

And for the first time, that resemblance felt negotiable.

None of this produced riots.

There were no crowds in the streets, no sudden rejection of institutions, no dramatic loss of order. The change was quieter and far more damaging to systems built on trust.

People pulled back.

They delayed commitments. They shortened horizons. They favored flexibility over permanence, liquidity over legacy. Why lock yourself into a future that institutions themselves now described as incomplete?

Withdrawal replaced outrage.

Withdrawal is rational and ruinous.

It wasn't rebellion.

It was recalibration.

It showed up as lower participation, softer engagement, a subtle reluctance to invest belief in long arcs. The social contract didn't shatter. It loosened.

That was the fear.

Not panic—recalibration.

Not that people would demand answers that didn't exist—but that they would stop believing answers mattered, or that explanation itself carried any obligation beyond performance. That once the public understood there was no fix, no deadline, no reassuring arc toward resolution, they would respond rationally.

By disengaging.

Institutions understood this instinctively, even when they couldn't articulate it cleanly. They had seen versions of it before—confidence leaving markets without a crash, trust eroding without scandal.

Disclosure threatened to make the future feel negotiable.

And negotiable futures are hard to govern.

You don't get riots.

You get people quietly stepping back from promises, from projections, from shared assumptions about where things are headed.

That kind of withdrawal doesn't break systems overnight.

It hollows them out.

Slowly is the point.

Slowly.

And once it begins, it is remarkably difficult to reverse.

∝ ∝ ∝

That was the calculus no one wanted to say out loud.

Once people internalized that some stabilizers were not permanent—that they weakened slowly, predictably, and without remedy—the future stopped feeling like a shared destination and started feeling like a negotiation each person conducted privately.

Institutions were built on the opposite assumption.

They depended on continuity being assumed rather than defended. On the idea that tomorrow would resemble today to plan around. When that assumption softened, coordination became harder than conflict. Shared timelines fractured into personal contingencies.

They relied on belief that long arcs made sense even when they were invisible. That pensions would mature. That infrastructure would outlive administrations. That education paid off because the world it prepared you for would still exist in roughly the same shape when you arrived.

Disclosure threatened that belief not by contradicting it outright, but by qualifying it.

And qualification is corrosive.

Because it never stops at one sentence.

A market can absorb bad news. It can even absorb shocks. What it struggles with is ambiguity that cannot be resolved on any actionable timeline. When uncertainty stretches beyond planning horizons, it stops being priced and starts being avoided.

Capital didn't flee.

It became cautious.

Long-term investments were reframed as philosophical positions rather than technical ones. Every projection began to sound like a statement of faith. Analysts still ran the numbers, but the confidence behind them thinned. Assumptions that once felt invisible had to be defended, and defending them exposed how much they rested on continuity no one could guarantee.

Governments felt it too.

Political promises lost depth. Voters stopped believing in the timelines attached to leaders. Ten-year plans sounded aspirational. Twenty-year plans sounded fictional. Anything beyond that felt almost impolite to propose.

The word *sustainable* became slippery.

It no longer meant enduring. It meant managed. Contested. Conditional. Every use invited follow-up questions that couldn't be answered without reopening the very uncertainty disclosure was meant to explain.

Sustainability stopped being a destination and became an argument.

None of this caused headlines.

It showed up as a change in tone. As a reluctance to commit. As a preference for reversible decisions. People still participated in institutions, but with less emotional leverage invested. They hedged not just financially, but psychologically.

Why believe deeply in futures described as incomplete?

That was the question institutions feared most— because it was reasonable.

Withdrawal didn't look like rebellion. It looked like quiet disengagement. Fewer people anchoring their lives to promises that assumed stability. More people keeping options open. More short-term thinking, not out of selfishness, but out of uncertainty about what long-term meant anymore.

The irony was painful.

Disclosure was resisted because it would provoke *clarity*—the kind that made people behave conservatively in ways systems depended on them not to.

Once the public understood that some risks were real, slow, and unfixed, optimism stopped being automatic. It had to be chosen.

And optimism that requires choice is no longer infrastructure.

It is effort.

Institutions sensed that effort wouldn't be evenly distributed. That some would continue believing, some would adapt, and some would quietly step back, taking their trust with them. No explosions. No collapse. Just a thinning of the collective willingness to commit to futures no one could confidently narrate.

That was why disclosure felt dangerous.

Because they wouldn't panic.

They would listen, understand just enough, and then decide—individually, rationally, silently—that the safest response was to pull away from long promises and live closer to the present.

You can govern panic.

Withdrawal has no negotiating table.

Withdrawal is harder.

It leaves no crowds to address, no demands to negotiate, no crisis to resolve. It leaves institutions standing, functional, intact—and gradually emptier of belief.

That was the risk Camp B was guarding against.

And the cost Camp A believed was already being paid.

Between them, the system hesitated.

Not frozen.

Just delaying.

Because as long as the future remained incomplete, so did responsibility for explaining what it meant to live inside it.

∝ ∝ ∝

THE COMPROMISE (HOW TRUTH LEAKS)

The compromise didn't arrive as a decision.

It arrived as default.

It arrived as a process.

No one stood up and said *this is how we'll handle it.* There was no vote, no formal resolution that split the difference between silence and disclosure. What emerged instead was a pattern—familiar, procedural, and comfortable enough that it barely registered as a choice.

Papers were written.

Carefully worded, densely cited, technically impeccable. They passed peer review without trouble because there nothing in them that was wrong. The data was presented cleanly. The uncertainty was acknowledged.

The implications were gestured toward and then bracketed off, left for "future work."

Panels were formed.

Interdisciplinary, balanced, meticulously representative. They convened at conferences and workshops with titles that promised scope and delivered restraint. The same questions were raised year after year, refined but unresolved. Recommendations were drafted, then softened, then deferred pending additional study.

Language adjusted itself.

Adjustment became doctrine without being named.

Drafts lost their sharp edges. Verbs weakened. Consequences became considerations. Statements that once implied direction were rephrased to emphasize duration. Nothing was removed outright. Everything was moderated.

In public summaries, the phrase *long-term orbital evolution* began to appear.

It sounded academic. Distant. Comfortably abstract. It suggested timeframes so vast they felt insulated from relevance. The words were accurate. They were also anesthetic.

Anesthesia is stability's method.

A reporter asked, "Does this affect anything people experience?"

"Not in daily life," came the answer.

Which was true.

No one connected the dots.

There was no line drawn between orbital mechanics and the clocks people relied on, the ecosystems that adjusted silently, the planning horizons institutions quietly revised. Each piece of information existed in isolation, accurate within its domain and inert outside it.

The truth was there.

Anyone motivated enough could find it—in journals, in appendices, in footnotes that gestured toward coupling and drift without ever assembling them into a story. Nothing was hidden. Nothing was denied.

But there was no narrative.

No explanation that moved from sky to system to life. No framing that said *this matters because it is slow*. No bridge between what was measured and what was lived.

The compromise satisfied everyone just enough.

Camp A could say the information was public.
Camp B could say it was responsibly contextualized.
Institutions could say they had been transparent.

And all of that was true.

What remained unspoken wasn't the data, but its consequence. Without connection, accuracy functioned as permission to ignore.

Ignorability became policy by omission.

What no one did was explain why these facts belonged together, or why their accumulation mattered more than any single result. The public received fragments—accurate, isolated, stripped of urgency by design.

The truth existed.

But without narrative, it floated.

That was the gap your novel steps into.

Not to invent facts.
Not to dramatize catastrophe.
But to assemble what already exists—quietly, carefully—into something coherent enough to be understood.

Because in the space between silence and disclosure, meaning doesn't vanish.

It migrates.

It waits.

∝ ∝ ∝

It waits in the margins.

In the space between what is said and what is connected. In the pause after a sentence ends cleanly and no one asks the obvious next question.

In the assumption that if something mattered to daily life, someone would have mentioned it plainly by now.

The compromise depends on that assumption.

As long as facts arrive detached—each accurate, each modest, each carefully bounded—they fail to cohere. The public encounters them the way it encounters background radiation: everywhere, measurable, and yet safely ignorable. It survives on fragmentation and procedural distance. A paper here. A panel there. A phrase that sounds important but distant enough to belong to someone else's concern.

Long-term orbital evolution.

It sounds like astronomy. It sounds like patience. It sounds like something that unfolds on timescales so large they cancel responsibility.

So nothing connects.

No one says that timekeeping debates and orbital drift share a cause. No one explains why ecosystems respond before cities do. No one draws a line from measurement to meaning, because drawing that line would require choosing a narrative—and narratives carry consequences.

The system prefers fragments.

Fragments are safe. They can be acknowledged without being owned. They can be cited without being synthesized. They allow institutions to say *we told you* later, without ever having said *this is what it means* now.

Even journalists, when they stumble across the material, struggle to assemble it.

"This feels… abstract," one says to an editor, scrolling through a draft. "There's no event."

"No villain," the editor replies. "No image. No timeline."

"So what's the angle?"

There isn't one that fits.

The story doesn't break. It accumulates. It doesn't accuse. It qualifies. It doesn't demand action. It asks for attention—and attention is the one thing modern systems ration carefully.

And so the compromise holds.

Truth circulates without momentum. Knowledge exists without consequence. Institutions remain intact. Confidence remains serviceable.

But something else begins to happen, quietly.

People who work closest to the data start speaking differently outside official settings. They choose metaphors instead of numbers. They test language on friends, on students, on anyone who will listen long enough to hear a connection rather than a fact.

They don't announce conclusions.

They tell stories.

Not dramatic ones. Not apocalyptic ones. Stories about systems that assume permanence and what happens when that assumption softens. Stories about slow change and why it matters precisely because it does not announce itself.

That is where the novel begins its work.

Not as revelation, but as assembly.

It does what reports cannot and panels will not. It takes fragments that already exist and places them in relation to one another. It restores scale where abstraction has flattened meaning. It translates without simplifying, connects without exaggerating.

It does not invent danger.

It names it.

And in doing so, it steps into the narrow space left open by the compromise—the space where truth has been released but never explained, where facts float free of narrative, waiting for someone to show how they belong to the same story.

That space is quiet.

But it is not empty.

And once something begins to fill it, the compromise can no longer pretend that saying less is the same as saying enough.

∝ ∝ ∝

Once that space is acknowledged, it changes how everything else sounds.

Old phrases lose their insulation. Familiar qualifiers stop feeling neutral. Words like *long-term* and *gradual* begin to register not as reassurances, but as descriptors that carry weight precisely because they refuse spectacle.

People start noticing patterns they were never meant to assemble.

A graduate student reads a paper on orbital dynamics and remembers a lecture on leap seconds. A coastal planner hears a radio segment on fisheries collapse and thinks about lunar timing charts she once dismissed as academic. A software engineer stumbles across a forum argument about time synchronization and realizes no one seems to agree on who owns the problem.

None of them panic.

They hesitate.

The hesitation is subtle, private, almost polite. It shows up as questions asked later than they should have been. As projects postponed without a clear reason. As an unease that doesn't attach itself to any single fact, but to the growing sense that the facts are no longer independent.

Institutions feel this first as friction.

Requests for clarification multiply. Briefings that once ended cleanly now run long because the room won't quite let go. Someone always asks one more question—usually framed as curiosity, sometimes as concern, never as accusation.

"How does this connect to…?"
"Is there a reason this keeps coming up across departments?"
"Are we tracking this somewhere holistically?"

The answers remain correct.

They are also incomplete.

And the gap between correctness and sufficiency becomes harder to ignore.

The compromise begins to strain—it succeeds too well. By keeping the truth dispersed, it allows meaning to form outside official channels.

People start doing synthesis on their own, without guidance, without shared language, without institutional pacing.

That is the part no one planned for.

A system can manage silence.
It can manage disclosure.
It struggles with distributed understanding.

Because once people assemble meaning independently, you no longer control the narrative's shape—only its accuracy. And accuracy alone does not determine trust.

This is where the novel operates most precisely.

Not as a counter-institution.
Not as an exposé.
But as a mirror held steady long enough for patterns to become undeniable.

It does not shout what reports whisper.
It does not resolve what panels defer.
It shows what happens when truths released without narrative begin to find one anyway.

Slowly.
Imperfectly.
Humanly.

It accumulates through inference rather than instruction.
Through conversations held without authority.
Through patterns noticed independently, then compared.
Meaning assembles sideways, not top-down.
Understanding arrives without permission.

The compromise was designed to prevent rupture.

What it couldn't prevent was interpretation.

And once interpretation begins—once facts stop floating and start clustering—the era defined by *technically accurate, but incomplete* enters its most unstable phase.

Not collapse.

Recognition.

The moment when enough people realize they have been holding pieces of the same story, and the only remaining question is who will say it out loud first—or whether the system will continue pretending that saying less is still the safest way to speak.

That moment does not announce itself either.

It arrives the way all the others have.

Quietly.

∝ ∝ ∝

Quiet does not mean still.

Once recognition begins, it spreads sideways rather than upward. It does not wait for permission. It moves through informal channels—conversations that start as jokes, as speculation, as *have you noticed…* and then don't quite resolve.

People begin to test the idea without naming it.

A planner mentions, casually, that long-lived infrastructure feels harder to justify than it used to. A professor reframes a lecture, not changing the content, just the emphasis. A journalist keeps a note open on her desktop labeled *patterns*, adding links she hasn't figured out how to connect yet.

None of this trips alarms.

That is why it works.

The compromise was built to manage statements. It was never designed to manage curiosity.

Institutions sense the shift only after it has already settled into the background. They notice it in tone rather than behavior. Questions sound less like requests for clarification and more like probes—gentle, respectful, persistent.

"Is this connected to anything else we're tracking?"

"Has this come up before in another context?"

"Are we assuming this stays abstract?"

The answers remain careful. The language remains calibrated. But something has changed in how it lands.

Incomplete no longer closes the conversation.

It postpones it.

People start to understand that postponement itself is a deliberate policy choice, not a condition imposed by ignorance. They realize that waiting for completeness is not a neutral act, but a preference for stability over synthesis.

That realization does not produce anger.

It produces distance.

Not withdrawal yet—just a subtle recalibration of trust. A sense that official narratives are accurate but insufficient, that they describe the pieces without acknowledging the shape those pieces form together.

This is where belief thins.

Not belief in science.

Not belief in institutions.

Belief in alignment.

The sense that knowledge flows cleanly from measurement to meaning begins to weaken. People start assuming that understanding requires effort on their part—that no one will assemble the story for them unless they do it themselves.

Some do.

Others don't.

The result is uneven comprehension, scattered insight, narratives that differ not in facts but in framing. Without a shared story, meaning fragments along personal lines: profession, temperament, tolerance for uncertainty.

The compromise holds formally.

Nothing forces it to break.

But informally, it stops containing what it was designed to manage. Truth no longer leaks as isolated facts. It leaks as intuition—harder to cite, harder to correct, harder to steer.

Institutions respond by refining language further.

More qualifiers.
More context without synthesis.
More emphasis on what is not known.

The irony is painful.

The more carefully they speak, the more people sense what is not being said.

That is the moment the novel inhabits.

Not the moment of revelation, but the long, ambiguous interval when everyone has access to the same information and no one agrees on what it adds up to. When silence is no longer total and disclosure is no longer sufficient.

The system is still intact.

But it is no longer narratively coherent.

And systems that lose narrative coherence don't fail immediately.

They drift.

They accumulate misunderstandings. They invite competing explanations. They create space—not for lies, but for stories that rush in to do the work official language refuses to do.

The compromise was meant to preserve order.

What it produces instead is a quiet multiplication of meanings.

And once meaning multiplies, control becomes an illusion.

∝ ∝ ∝

The next shift is almost imperceptible.

It happens when people stop asking institutions for meaning at all.

Not out of rebellion. Out of efficiency.

If official language is going to remain technically correct and narratively empty, then interpretation becomes a private act. People stop waiting for synthesis and begin constructing their own, using whatever fragments they have access to and whatever frameworks feel trustworthy.

This is not misinformation.

It is unsupervised sense-making.

A systems engineer frames the issue as a maintenance problem. A philosopher frames it as a limit to human planning. An economist reframes long-term risk as a discounting failure. An environmentalist folds it into existing climate narratives. None of these are wrong. None of them are complete.

They coexist.

The compromise never anticipated plurality. It assumed that if nothing dramatic was said, nothing dramatic would be understood. It underestimated how well humans detect pattern once they start looking for it.

Institutions notice this only indirectly.

They see it in misaligned questions. In briefings where attendees seem to be having a different conversation than the one on the screen. In public comments that aren't hostile, just… sideways.

"That's not what we meant," someone says.

And the reply comes back, politely, "Then what *did* you mean?"

That question is harder to answer than it should be.

Because the compromise has hollowed out intention. Official statements are no longer designed to convey meaning—only to avoid consequence. They describe, qualify, defer. They don't orient.

Orientation is left to the listener.

At first, institutions try to correct interpretations piecemeal.

Clarifications are issued. FAQs are added. Footnotes grow. None of it restores coherence, because coherence requires choosing a story, and choosing a story requires accepting responsibility for how it lands.

No one wants that responsibility.

So the system doubles down on process.

More panels. More listening sessions. More stakeholder engagement that produces insight but no synthesis. Everyone is invited to speak. No one is authorized to connect.

The result is not silence.

It's noise with no center.

This is where the compromise finally shows its weakness.

By refusing to tell a story, it does not prevent stories from forming. It ensures they form unevenly, without shared language, without agreed scope, without a common sense of scale.

And once that happens, trust doesn't collapse.

It fragments.

People don't stop believing facts.

They stop believing that facts will be *assembled for them.*

That is the vacuum the novel fills.

Not by shouting what institutions whisper, but by doing the one thing the compromise will not: placing facts in relation and allowing implications to exist without immediately softening them.

The novel doesn't resolve the uncertainty.

It respects it.

But it refuses to pretend that uncertainty means meaninglessness.

Because the final truth the compromise cannot admit is this:

When truth exists without narrative, someone will supply one.

The only question is whether that story is careful—or convenient.

Institutions chose careful silence.

The world chose interpretation.

And interpretation, once it spreads, does not wait for permission to become real.

The compromise also changes who feels responsible.

When truth is fragmented, responsibility fragments with it. No single office owns the implications. No spokesperson carries the weight of synthesis or the burden of drawing conclusions. Everyone can point to process—peer review, panels, summaries—as evidence that the system is functioning.

And it is.

Just not in the way people assume.

Responsibility dissolves into roles. Analysts analyze. Communicators communicate. Policymakers wait for something actionable. Each piece performs its function correctly, and the whole avoids ever having to answer the simplest question: *What does this mean for how we live now?*

That question becomes radioactive.

It has no single owner. Answering it would require crossing boundaries—scientific into social, technical into ethical, descriptive into normative. It would require someone to say *this matters* without being able to say *and here's what you do.*

So no one does.

Instead, the system learns to move laterally around meaning. It produces ever more precise descriptions of parts, trusting that the whole will remain someone else's problem. The compromise becomes self-sustaining because it allows everyone to remain professionally correct.

Over time, this creates a quiet inversion.

The public is told that nothing connects to daily life. Yet daily life begins to feel subtly unmoored.

People cannot name the cause. They only sense that long promises feel thinner than they used to, that stability now carries asterisks, that confidence requires more effort than it once did. They are not reacting to a fact. They are reacting to an absence of explanation.

The system interprets this as unrelated.

Consumer sentiment. Political volatility. Cultural fatigue.

Everything but synthesis.

That is the final irony of the compromise: by refusing to tell a story in order to preserve stability, it allows instability to appear everywhere except where it originates. Effects are noticed without causes. Drift is felt without direction.

The truth exists.

The data is solid.

The language is careful.

What's missing is permission—to connect, to contextualize, to admit that slow change still shapes how societies orient themselves toward the future.

That permission never comes from within the system.

So it comes from outside.

Not as policy.
Not as briefing.
As narrative.

That is where the novel ends its work in this block.

Not by breaking the compromise, but by showing what it costs.

Not by exposing a lie, but by assembling truths that were always there—and letting them finally stand in the same frame long enough to be seen as part of the same world.

PHASE 6 — THE LONG VIEW: WHAT MILLIONS OF YEARS ACTUALLY DO

THE MOON'S OUTWARD DRIFT IS NOT A THEORY

The tone changes here.

Not because anything new is discovered, and not because the facts become more dramatic. The change comes from scale. From stepping far enough back that urgency loses its grip and something quieter, heavier, takes its place.

This is the long view.

It trades urgency for weight.

It does not predict catastrophe. It does not count down. It does not argue for action or restraint. It explains what happens when systems governed by simple, relentless physics are allowed to run for spans of time far longer than human memory, politics, or narrative comfort.

From this distance, panic looks misplaced. So does reassurance.

What remains is inevitability—not as threat, but as condition.

The Moon's outward drift is not a hypothesis waiting for confirmation. It is not a contested model or a speculative projection. It is a consequence of mechanics so ordinary they are taught early, then forgotten because they seem too simple to matter.

Earth spins.

The Moon pulls.

The oceans respond.

Energy moves.

Angular momentum is transferred. Slowly. Continuously. From Earth's rotation into the Moon's orbit. Earth loses a trace of spin. The Moon gains a trace of distance. The exchange is not dramatic. It does not surge or stall. It follows the same rules today that it followed billions of years ago.

This is not debated.

Debate ended when the measurements stopped disagreeing.

It is accepted to the point of invisibility.

Agreement has rendered it background.

The mathematics is settled. The observations agree. The fossil record of tides, written into ancient sediments and growth rings, confirms the story in layers laid down long before humans existed to measure them. Days were shorter. The Moon was closer. The rhythm was faster.

Time lengthened.

Distance accumulated.

Nothing intervened.

The process does not require intention. It does not care about habitability, calendars, or meaning. It is not responsive to concern. Unless physics itself changes, it will continue because nothing stops it.

This is where the discomfort begins.

From the long view, stability is not a fixed state. It is a moving balance maintained by forces that don't negotiate. Earth's rotation is not preserved. It is spent. The Moon's orbit is not static. It evolves. The system does not seek equilibrium in a way humans would recognize as rest or relaxation.

It seeks conservation.

What feels stable across a lifetime is revealed, across millions of years, as a slow exchange that never quite finishes. Continents drift. Axes wobble. Orbits widen. Nothing snaps. Nothing announces a threshold crossed.

The unsettling part is not that change happens.

It is that change happens without regard to whether anyone is prepared to interpret it.

When this is explained plainly, without hedging or metaphor, it strips away certain comforts. Not the comfort of safety, but the comfort of finality. There is no point at which someone stands up and says *this is when it stops.* There is no stable configuration the system is working toward that coincides neatly with human needs.

There is only motion.

Measured motion. Predictable motion. Motion governed by rules so well understood that disagreement about them would be an error, not a debate.

This is what makes the long view feel calm in a way that borders on indifference. The system does not accelerate to meet attention. It does not slow down to accommodate concern. It continues at the same pace it always has, indifferent to who is watching and how precisely they measure.

For most of human history, this didn't matter. The changes unfolded too slowly to intrude on planning horizons that rarely extended beyond a few generations. The Moon's retreat was invisible. The lengthening of the day was imperceptible. The system could be treated as effectively constant because, for all practical purposes, it was.

∝ ∝ ∝

But once measurement entered the picture—once light itself was timed, once milliseconds mattered, once models stretched forward rather than backward—the long view stopped being abstract.

It became operational.

Abstraction ended where coordination began.

This block does not ask what happens next year, or next decade, or even next century. It asks what happens when a civilization realizes that the forces underwriting its sense of permanence have never been permanent at all—and never pretended to be.

Nothing here is controversial.

Nothing here requires belief.

The Moon moves outward.

Earth's rotation slows.

Days lengthen.

This has been happening for billions of years.

It will continue.

The calm comes from accepting that this is not a problem waiting to be solved.

It is a condition waiting to be understood.

And understanding it—not urgently, not defensively, but honestly—changes the frame through which everything else must eventually be seen.

∝ ∝ ∝

The difficulty is not understanding the physics.

The difficulty is accepting how little it cares about scale.

Human reasoning is tuned to thresholds. To points of failure. To moments when a system either holds or breaks. We are trained—by stories, by engineering, by politics—to look for stress limits, warning signs, deadlines.

The long view offers none of these.

It replaces thresholds with accumulation. It replaces deadlines with gradients. It replaces failure with drift so gradual that no single moment deserves blame or credit.

From this perspective, the Moon's retreat is not a change *toward* anything. It is simply motion continuing. The system does not remember where it started and does not anticipate where it ends. It only conserves what it must and dissipates the rest.

This is where inevitability stops feeling dramatic and starts feeling ordinary.

Familiarity replaces urgency.
Continuity masquerades as permanence.
Change proceeds without asking to be noticed.

The Earth–Moon system is not unstable. It is not fragile. It is not approaching collapse. It is behaving exactly as systems governed by conservation laws behave when left alone for billions of years.

Which is to say: it evolves.

Deep time does not rush. It does not build tension. It does not reward attention. It simply allows small effects to finish what they begin.

A few milliseconds per century become hours per rotation. A few centimeters per year become vast orbital separations. Nothing dramatic happens along the way.

And that is the point.

From millions of years out, catastrophe is rare. What dominates, instead, is continuity under slow transformation. The system remains recognizable long after its parameters no longer resemble their original values.

This is deeply unintuitive to creatures who live inside stories.

We expect change to announce itself. We expect importance to correlate with immediacy. We expect systems to justify themselves by responding when noticed.

The long view violates all of that.

It insists that some of the most consequential processes are the least responsive to attention. That some outcomes become unavoidable because nothing interrupted them.

This is why inevitability feels unsettling only when framed emotionally.

Stripped of drama, it becomes almost peaceful.

There is no race to intervene. No moral urgency embedded in the physics. No suggestion that anything has gone wrong. The system does not need saving from itself.

What *does* struggle in this frame is the human habit of assuming permanence unless warned otherwise.

Deep time does not warn.

It only proceeds.

And once that is understood—not abstractly, but viscerally—the question shifts again. Not *what will happen?* and not *how do we stop it?* but something quieter and more difficult to hold:

What does it mean to anchor meaning, stability, and promise inside processes that are indifferent to them?

That question does not belong to catastrophe.

It belongs to inevitability.

∝ ∝ ∝

WHAT HAPPENS *FIRST* (LONG BEFORE THE MOON IS "GONE")

The first thing that changes is not distance.

It is balance.

Long before the Moon is ever meaningfully "gone," long before it becomes a different object in the sky, something subtler begins to loosen. The effect does not announce itself with violence or symmetry. It does not arrive on a schedule that can be marked or forecast cleanly.

It arrives as sensitivity.

Sensitivity precedes instability.

Earth does not spin upright by accident. Its axial tilt—the angle that gives rise to seasons, climates, long-term habitability—is not fixed by inertia alone. It is held in check by a stabilizing influence that never touches the surface directly: the Moon's gravity acting as a damper on chaos.

Without that influence, Earth's axis wouldn't snap or topple. It would wander.

The Moon's presence smooths that wandering. It resists rapid swings. It suppresses resonance. It turns what would otherwise be erratic motion into something slow, bounded, and familiar. Seasons repeat. Climate regimes persist long enough for ecosystems to adapt and civilizations to assume continuity.

As the Moon recedes, that stabilizing torque weakens.

Not abruptly.

Not cleanly.

And not in a way that can be summarized as a single trend line.

This is where intuition fails.

People imagine loss of stability as collapse—a sudden increase in tilt, a dramatic lurch. But that is not how coupled gravitational systems behave. What weakens first is not the average state, but resistance to disturbance.

Earth's axial tilt becomes more sensitive.

Sensitive to gravitational nudges from other planets.
Sensitive to internal redistributions of mass.
Sensitive to long-period resonances that were once damped into irrelevance.

The axis does not "break." It listens more closely.

This sensitivity does not produce immediate chaos. For long stretches, nothing looks different at all. The tilt remains near its current value. Seasons still feel recognizable. Climate still oscillates within ranges that seem normal by historical standards.

That is precisely what makes this effect so important.

Because once damping weakens, predictability weakens with it.

The system becomes harder to bound. Variability increases. Small perturbations that once decayed now persist longer. Some are amplified. Others interact in ways that cannot be reduced to clean cycles.

The axis begins to behave less like a metronome and more like a pendulum in a room where the air itself is moving.

No single moment marks the transition.

There is no "loss of axial stabilization" event.

There is only a gradual shift from certainty to contingency.

Contingency is stability without guarantees.

This happens long before any dramatic change in distance matters. Long before tides vanish or the Moon shrinks to a point of light. Long before anything feels existential.

It is the earliest meaningful consequence because it operates upstream of everything else.

Climate sensitivity.
Seasonal variability.
Long-term habitability windows.

All of these depend not just on where the axis points, but on how reliably it stays there.

As the Moon moves outward, Earth does not lose its tilt.

It loses some of its protection against wandering.

Not suddenly.

Not predictably.

But inexorably.

And that makes this the most important effect—it quietly changes the character of stability itself.

From something enforced

to something negotiated.

Over spans of time measured not in lifetimes or civilizations, but in the slow patience of physics.

This is what happens first.

And once it begins, nothing dramatic needs to follow for the consequences to matter.

Once that protection thins, the meaning of *stable* begins to change.

Not in any way a calendar would capture. Not in any way a single generation would experience as loss. The axis does not announce its new sensitivity. It behaves almost exactly as it always has—until it doesn't, and even then the difference is subtle enough to be arguable.

From the long view, what weakens first is not control, but forgiveness.

Perturbations that once damped out over tens of thousands of years linger longer. Oscillations that once stayed bounded acquire wider envelopes. The system remains coherent, but it becomes less disciplined. Its future states spread out, not wildly, but enough that clean extrapolation grows harder.

This is why prediction fails before experience does.

Models still run. Projections still converge—at first. But the confidence intervals widen. The neatness erodes. The farther forward one looks, the less certain it becomes that the axis will remain close to where it is now, or that it will wander only within the ranges life has already adapted to.

Nothing forces a dramatic outcome.

That is the unsettling part.

The loss of axial stabilization does not push Earth toward a single fate. It opens a space of possible behaviors. Some are mild. Some are disruptive. Which ones occur depends on interactions that are individually small and collectively decisive.

Jupiter's pull.
Mars's resonance.
The redistribution of mass on Earth itself.

With the Moon close, these influences are muted. With the Moon farther away, they speak more loudly—not enough to shout, but enough to be heard.

Over deep time, that matters more than distance ever could.

Axial tilt governs how sunlight is distributed across the planet. It shapes ice ages, deserts, monsoons, and migration corridors. When tilt varies slowly and predictably, ecosystems adapt. When it becomes more variable, adaptation has to work harder and fails more often.

Again, nothing collapses.

But the margins thin.

Periods of relative stability shorten. Transitions take longer to settle. Climate regimes overlap instead of replacing one another cleanly. The planet does not become hostile. It becomes less forgiving of assumption.

This is why the effect precedes everything else.

Before oceans change their reach.
Before days grow noticeably longer.
Before the Moon's distance becomes visually meaningful.

The planet's posture toward the Sun becomes less fixed.

And posture, more than position, determines habitability.

From a human perspective, this is almost impossible to feel. From a geological perspective, it is immediate. From the long view, it is decisive.

The Moon does not need to leave for its role to diminish.

It only needs to move far enough that its steadying hand lightens.

And once that happens, Earth is not thrown into chaos.

It is released into possibility.

Some of those possibilities are benign. Others are not. None of them require catastrophe to matter.

This is what makes the long view so difficult to hold.

The future does not arrive as a verdict.

It arrives as variance.

And variance, once introduced, does not announce how far it will spread—only that the system will never again behave quite as narrowly as it once did.

That is what happens first.

Everything else follows from that widening.

∝ ∝ ∝

The loss of axial stabilization does not show itself where people expect to see change.

It does not appear as a new season.
It does not arrive as a new climate.
It does not announce itself as disorder.

It appears as a slow erosion of *repeatability*.

The same inputs no longer guarantee the same outcomes. Patterns that once returned on schedule begin to slip. Not enough to invalidate them—just enough to complicate them without overturning conclusions. Enough to unsettle confidence without replacement. The past becomes a less reliable guide to the future. The system no longer promises to behave as narrowly as it once did.

This is invisible on short records.

If you measure a century, you see noise.
If you measure a millennium, you see drift.
If you measure tens of thousands of years, you begin to see something

else: a system that no longer snaps back to familiar configurations as readily.

The Moon's stabilizing effect was never about locking Earth into a single tilt. It was about constraining the range of motion. About keeping the planet's orientation within corridors that life learned to navigate.

As that constraint weakens, those corridors widen.

The widening does not prefer one direction. It does not "aim" at disaster. It simply allows excursions that were once unlikely to become merely rare. And rare events, given enough time, become inevitable.

This is where the long view becomes uncomfortable.

It refuses to guarantee any particular future.

With strong axial damping, Earth's tilt behaves like a well-trained system: perturbed, then corrected. With weaker damping, correction still occurs, but more slowly, and sometimes incompletely. Perturbations overlap. Memory accumulates.

The system begins to remember disturbances instead of forgetting them.

From that point on, history matters more.

Small differences compound. Initial conditions matter longer in systems without persistent intervention. The planet's trajectory becomes more contingent on its past rather than constantly re-centered by a dominant stabilizer.

That is the quiet shift.

Not chaos.

Not collapse.

Contingency.

This is why the effect precedes everything else in importance. Distance can be measured. Day length can be counted. Tides can be modeled. But contingency resists narration. It shows up only as increasing difficulty in making confident claims about what comes next.

It does not demand action.

It demands humility.

From the long view, the Moon's retreat is not a countdown to absence. It is a gradual withdrawal of certainty. The planet does not lose its seasons. It loses some of the guarantee that they will always resemble the ones that life had evolved to expect.

That guarantee has never been explicit.

It was simply enforced by physics.

As that enforcement weakens, nothing dramatic replaces it. There is no new stabilizer waiting. No alternative mechanism steps in to restore the old constraints.

The system does not fail; it simply continues without the same counterbalance.

What had once been held in check is allowed to drift, governed by the same rules as before, but with fewer restraints shaping the outcome.

There is only a wider field of possible futures.

And once that field opens, it never closes again.

That is why this happens first.

Because everything else—climate, habitability, even the meaning of "long-term stability"—depends less on where Earth is, and more on how tightly its behavior is bound.

The Moon does not need to disappear to change that.

It only needs to continue doing what it has always done.

Move away.

∝ ∝ ∝

What Instability Really Means

Axial tilt is not an abstract parameter.

It is the mechanism by which sunlight is distributed across the planet. It determines where seasons exist at all, how intense they are, and how reliably they return. It governs whether ice accumulates slowly or collapses suddenly, whether deserts stay confined to latitudes that learned how to endure them, whether ecosystems migrate gradually or are forced to retreat in pulses.

Tilt is not weather.

It is climate's architecture.

When Earth's axial tilt is well constrained—held within narrow bounds and corrected gently when perturbed—seasons repeat with a familiarity life can plan around. Ice ages arrive and retreat on long, intelligible cycles. Habitable zones remain where biology has already learned how to function.

This is because the climate rules are consistent.

As lunar stabilization weakens, those rules loosen.

The average tilt may remain near where it is now for long spans of time, but the range around that average widens. The axis wanders more freely. It responds more strongly to planetary perturbations that once barely registered. Resonances that were previously damped into irrelevance gain leverage.

Nothing explodes.

Instead, climate behavior becomes harder to bound.

Over millions of years, this manifests not as a single dramatic shift, but as a sequence of regimes that don't repeat cleanly. Ice ages deepen beyond familiar limits, then retreat faster than ecosystems expect. Glacial periods overlap. Interglacials shorten or stretch unpredictably.

Deserts don't simply expand.

They migrate.

Dry zones creep toward poles, linger, then withdraw. Tropical belts shift north or south without settling into stable corridors. Regions that once experienced gradual change find themselves oscillating between extremes without enough time to adapt fully to either.

This is not unprecedented in the solar system.

Mars offers the boundary condition—not as a warning, but as an example of what happens when axial tilt is allowed to roam without a strong stabilizing companion. Mars does not lose its atmosphere in moments. It does not collapse into sterility overnight. It cycles through wildly different climate states over geological time, driven in large part by uncontrolled axial variation.

Earth does not become Mars.

But it becomes less constrained.

The important distinction is not survivability.

Life is resilient. It adapts, migrates, collapses locally, re-emerges elsewhere. Given time, it fills whatever niches remain accessible. A planet does not need stable seasons to host biology.

Civilization does.

Civilization depends on repeatability. On agricultural zones that persist long enough to justify infrastructure. On coastlines that don't wander faster than cities can respond. On climate regimes that allow planning to extend beyond the next few generations.

Axial instability does not remove habitability.

It erodes predictability.

And predictability is the substrate civilization is built on.

Under widening axial variance, there is no single climate to adapt to—only a moving target. Adaptation becomes reactive rather than anticipatory. Long-term planning becomes speculative. The concept of "normal" loses coherence.

Nothing here implies extinction.

It implies friction.

More failed adaptations.

More abandoned regions.

More cultural compression around stable intervals that don't last.

Life survives.

It always has.

Civilization struggles because the planet becomes less legible. The rules don't disappear. They stop repeating cleanly enough to anchor long promises.

This is the mistake apocalypse stories make.

They look for endings.

The long view offers something quieter: a planet that continues to function, continues to host life, continues to change—while the structures that depend on consistency find themselves increasingly out of phase with the world beneath them.

That is what axial instability actually means.

Not collapse.

But a future that refuses to stay still long enough to be inherited unchanged.

∝ ∝ ∝

What makes this especially difficult to accept is that nothing here violates intuition at the level of experience.

A civilization can live for centuries inside a climate regime and never realize how narrow that regime is. The boundaries feel generous because they are invisible. Crops fail and recover. Ice advances and retreats. Storms come and go. Everything that happens can be explained locally, without reference to deep structure.

Axial instability does not announce itself as a cause.

It expresses itself as coincidence.

A century of harsher winters overlaps with a century of agricultural stress. A period of rapid glacial retreat aligns with coastal disruption and population movement. None of these events require a celestial explanation to make sense on their own. Each has proximate causes that satisfy inquiry. Temperature. Precipitation. Ocean circulation. Volcanism. Human activity.

The long view does not deny those causes.

It reframes their reliability.

As axial variation increases, climate drivers stop reinforcing each other predictably. Feedbacks misalign. Patterns that once offset extremes begin to stack them instead. The system does not become chaotic in the sense of random—but it becomes sensitive enough that small changes propagate farther than expected.

This is where civilization encounters its limits.

Civilization is not merely the presence of people or tools. It is an agreement with the future. A promise that investment made now will still make sense later. That roads will connect the same places. That rivers will flow within known bounds. That seasons will remain close enough to their historical behavior to justify planning beyond a single lifetime.

Axial instability does not revoke that agreement outright.

It revises the terms without asking.

The cost appears first in adaptation fatigue. Each generation is forced to adjust to conditions slightly different from the last, without ever stabilizing long enough for the adjustment to feel complete. Solutions become temporary by default. Resilience strategies are revised before they can mature.

This is not collapse.

It is attrition.

Regions that once supported dense populations thin out because they become unreliable. Agriculture migrates. Infrastructure is abandoned not in ruins, but in quiet disuse. Cultural memory stretches to explain changes that no longer fit inherited stories.

What looks, from the inside, like political failure or economic mismanagement is often something deeper: the environment no longer behaving as consistently as the systems built upon it require.

The planet remains generous.

Just less predictable.

From the long view, this is not a tragedy. It is a transition between modes of habitability. Earth does not cease to support life. It ceases to support *long continuity without renegotiation.*

And renegotiation is expensive.

It requires constant attention, constant revision, constant willingness to abandon structures that once worked well. It demands flexibility from institutions designed around permanence.

That mismatch is the pressure civilization feels.

Not a final crisis.

A permanent one.

This is why axial instability matters long before any dramatic change in distance, tides, or day length. It reaches into the assumptions beneath planning itself. It turns stability from a background condition into an active variable—something that must be monitored, managed, and never fully trusted.

Life does not care about that distinction.

It adapts.

Civilization does care.

Because civilization is not adaptation alone.

It is memory.

And memory depends on repetition.

When repetition weakens, the future does not end.

It becomes harder to recognize.

That is the real consequence of axial instability.

Not extinction.

But a world that insists, over and over again, that nothing built on yesterday's expectations is guaranteed to remain aligned with tomorrow's ground.

∝ ∝ ∝

Axial instability does not just alter climate.

It alters **inheritance**.

Civilizations don't merely adapt to environments. They inherit them. Knowledge is passed down with the assumption that the world it describes will still resemble itself closely enough to make that knowledge useful. Calendars are inherited. Planting seasons are inherited. Legal boundaries, water rights, trade routes, even myths—all of them encode expectations about where heat, cold, rain, and scarcity will tend to reside.

Axial stability makes that inheritance possible.

When the tilt behaves predictably, memory remains valuable. The past continues to explain the future well enough that experience compounds rather than expires. Cultures can build depth instead of constantly reorienting.

As axial variance widens, that inheritance degrades.

Not dramatically. Quietly.

The Sun is not passive in this.

It does not cause the instability, but it is the reference against which all of it is measured.

Axial tilt only matters because there is a star at the center of the system. Seasons exist because Earth's axis is inclined relative to its orbit around the Sun. Climate patterns emerge because solar energy is distributed unevenly across latitudes, concentrated or diluted depending on where the planet leans.

The Sun provides the gradient.

The Moon moderates how that gradient moves.

As long as Earth's axial orientation remains tightly constrained, the relationship between planet and star stays legible. Insolation patterns repeat. The same regions receive roughly the same annual energy, year after year, epoch after epoch. The Sun does not change its behavior, and Earth presents itself to it in familiar ways.

Axial instability breaks that predictability.

Not by altering the Sun, but by altering how Earth *faces* it.

As lunar stabilization weakens, Earth's axis becomes more responsive to forces that were once damped. The Sun's gravity, steady and immense, participates automatically in this sensitivity. It does not need to vary. Its constancy becomes part of the problem.

Small changes in axial orientation translate directly into large redistributions of solar energy. A few degrees of additional tilt shift the balance of heating across all continents and oceans. Polar regions receive more sunlight during parts of the year, accelerating ice loss asymmetrically.

Mid-latitudes experience more extreme seasonal contrasts. Equatorial zones widen or narrow unpredictably.

The Sun amplifies whatever the axis allows.

This is why axial instability matters more than orbital distance in the near deep-time sense. Earth can move slightly closer or farther from the Sun and remain habitable. But how Earth presents itself to that energy source—how steadily, how predictably—determines whether climate regimes repeat cleanly or drift into new configurations.

The Sun does not destabilize Earth.

It enforces the consequences of instability.

Once axial variance increases, the Sun's energy distribution becomes a multiplier. Differences that once averaged out now persist long enough to reshape ice sheets, alter circulation patterns, and redraw climate belts.

From the long view, the Sun is the metronome.

The Moon was the damper.

As the damper weakens, the metronome does not slow or change tempo. It continues, relentlessly consistent, exposing every wobble the system can no longer suppress.

This is why the system remains calm even as it becomes less forgiving. Nothing external attacks the Earth. No new force is introduced. The Sun shines as it always has. Gravity behaves as expected.

What changes is Earth's ability to keep presenting the same face to the same energy source.

That relationship—planet to star, posture to light—is what defines climate over geological time. And once posture becomes more variable, the Sun ensures that variability has consequences.

Not suddenly.

Not predictably.

But thoroughly.

The Moon's retreat does not act alone. It loosens Earth's grip on its own orientation. The Sun then does what it has always done: it delivers energy according to geometry, not intention.

Life adapts to that.

Civilization, built on the assumption that the Sun will keep arriving in familiar ways, finds the adjustment harder.

Earth, over deep time, stops promising to meet it the same way twice.

Traditional knowledge begins to fail first—its assumptions no longer repeat reliably. Farmers notice that old rules of thumb still work sometimes, then fail without pattern. Water management systems built for historical melt cycles face decades where they underperform, then decades where they are overwhelmed.

Nothing forces an admission that the system has changed.

People blame governance. Or mismanagement. Or bad luck.

Civilization responds the way it always does: by adding layers. More regulation. More monitoring. More contingency. The system grows heavier in response to instability rather than lighter.

This creates a second-order effect.

The cost of continuity rises.

Maintaining civilization in a less predictable climate does not require heroism or collapse. It requires **constant correction**, applied continuously and without final resolution. More resources devoted to forecasting. More capital tied up in resilience that may never be used. More political energy spent arbitrating between regions that no longer share the same environmental rhythms.

The planet still supports life easily.

It simply stops subsidizing civilization's assumptions.

That subsidy—predictable axial behavior over long spans—has always been invisible. It was never earned. It was inherited from a configuration that happened to be stable enough, long enough, for complexity to accumulate.

Axial instability withdraws that subsidy slowly.

The result is not a world that cannot be lived in.

It is a world that must be continually renegotiated.

Borders become provisional. Food systems regionalize, then fragment. Cultural coherence weakens because shared experience thins. The future stops looking like a continuation of the past and starts looking like a sequence of local adjustments.

From the inside, this feels like decline.

From the long view, it is transition.

Earth does not become hostile.

It becomes less cooperative.

Life thrives in that. It always has. Evolution loves variability. It exploits it.

Civilization does not.

Civilization depends on compressing variability into norms—on treating the future as something that can be planned rather than constantly re-learned.

Axial instability does not end that project.

It makes it more expensive than any civilization has ever had to budget for.

And because the change unfolds across millions of years, no single society ever experiences the full arc. Each one encounters only a segment—a period where the past explains less than it used to, and the future requires more effort to imagine.

That is why this point matters.

Because it explains why, over deep time, civilizations appear and vanish even while life continues uninterrupted. Not through catastrophe, but through the slow erosion of the conditions that make continuity easy.

Axial instability does not kill worlds.

It makes them harder to remember.

And a civilization that cannot rely on memory repeating itself eventually becomes something else—not destroyed, not defeated, just unable to carry its former shape forward.

That is still not apocalypse.

It is inevitability, doing what it always does. Quietly.

∝ ∝ ∝

What Happens To Tides And Oceans

The next changes don't arrive from the sky.

They arrive at the shoreline.

Tides are among the most familiar rhythms on Earth. They feel ancient, reliable, almost comforting in their regularity. Water rises. Water falls. Coasts breathe. Harbors empty and refill. Nothing about this seems fragile.

That familiarity hides how active tides really are.

They are not just motion. They are work.

Work once provided without accounting.

As the Moon recedes, its gravitational pull on Earth weakens—not enough to erase tides, not enough to flatten oceans into stillness, but enough to reduce their reach and their force. The difference is not something a person standing on a beach would notice. The water still comes in. The water still goes out.

What changes first is not height.

It is mixing.

Strong tides stir coastal waters, resuspend nutrients, oxygenate shallows, and connect nearshore ecosystems to the open ocean. They act as a slow but persistent engine, moving energy and matter between regions that would otherwise remain isolated.

As tidal forcing weakens, that engine loses efficiency.

Coastal waters stratify more easily. Nutrients settle where they once circulated. Oxygen gradients sharpen. Some regions become quieter, chemically speaking—not calmer, but less connected.

This does not kill the ocean.

It reorganizes it.

Marine ecosystems are built around motion. Plankton blooms depend on nutrient upwelling. Fisheries depend on the timing and location of those blooms. Coral systems depend on flushing that removes waste and brings food. Intertidal zones depend on exposure cycles that select for resilience and flexibility.

When tidal energy diminishes, these systems don't fail all at once.

Change distributes itself unevenly across regions and species. Stability erodes locally long before it registers globally.

They shift.

Species that thrive in highly mixed environments lose ground. Others, adapted to calmer, more stratified waters, expand. Food webs reconfigure. Productivity migrates rather than disappears.

From a biological perspective, this is adaptation.

From a civilizational perspective, it is disruption.

Fishing grounds that once returned year after year begin to wander. Coastal ecologies that supported predictable harvests become intermittent. The problem is not scarcity—it is reliability.

Again, nothing dramatic announces the change.

There are no great floods. No sudden drownings of coastlines. No cinematic catastrophe. Sea level rise, where it occurs, has other drivers and other stories.

This is not extinction by flood.

It is extinction by reconfiguration.

The ocean remains vast. Life remains abundant. But the patterns humans learned to depend on—seasonal runs, spawning grounds, nutrient pulses—become less stable over long spans of time.

Civilizations that oriented themselves around coasts did so because coasts were interfaces: between land and sea, between predictability and abundance. Tides were part of what made those interfaces productive. They ensured constant exchange.

As that exchange weakens, coastlines don't vanish.

They become quieter in some places, more chemically extreme in others, and less uniformly generous overall.

This matters because oceans are not just ecosystems.

They are regulators.

They absorb heat. They buffer climate. They move energy across latitudes. Changes in coastal mixing ripple outward, subtly altering circulation patterns that influence weather far inland.

Once again, the effect is not a single outcome.

It is a redistribution of conditions.

Some regions benefit. Others lose. The global system remains functional. But its behavior becomes less familiar, less repeatable, and harder to plan around.

Life adapts.

It always does.

But civilizations that grew around stable coastal productivity find themselves renegotiating their relationship with the sea—because it stopped behaving quite the way it always had.

The Moon does not need to disappear to cause this.

It only needs to continue moving away.

And as with everything else in the long view, the ocean does not protest.

It simply adjusts—slowly, patiently, and without regard for who learned its rhythms when they were stronger.

∝ ∝ ∝

Tides don't only move water.

They move time.

Coastal life is synchronized to tidal cycles with a precision that rivals circadian rhythms. Spawning events, feeding windows, migrations, and predator–prey interactions are timed not just to day and night, but to the rise and fall of the sea. These cycles provide a metronome—slow, predictable, and deeply embedded.

As tidal forces weaken, that metronome softens.

The timing does not vanish. It blurs.

High and low tides still occur, but with reduced contrast. The cues organisms rely on become less decisive. Some species adjust easily.

Others misfire—spawning too early, feeding too late, missing narrow windows that once repeated reliably.

This does not register as extinction at first.

It registers as inefficiency.

Lower reproductive success here. A missed cohort there. A food web that still exists, but transfers energy less cleanly from one level to the next, with small losses accumulating over time. Over long spans, inefficiency accumulates into restructuring, slow enough to evade notice, consistent enough to endure.

The same is true for human systems layered on top.

Ports are engineered around tidal ranges. Shipping schedules, dredging cycles, harbor designs—all assume a certain energy budget provided for free by the Moon. As that budget shrinks, maintenance costs rise. Interventions that were once occasional become permanent. Systems designed for assistance become dependent on it.

Nothing breaks.

It just stops being self-sustaining.

This is a recurring pattern in the long view: natural processes that once subsidized civilization slowly withdraw that subsidy. Not out of hostility. Out of physics.

Tides were never guaranteed.

They were inherited.

As their strength diminishes, oceans don't become still. But the relationship between land and sea becomes less animated, less forgiving of neglect. Regions that once relied on natural flushing must replace it artificially—or accept degradation.

Again, this is not apocalypse.

It is accounting.

Energy that was once provided by orbital mechanics must be supplied elsewhere if the same outcomes are desired. Many places will choose not to pay that cost. They will adapt differently instead.

Coastal civilizations don't vanish.

They reorient.

Some retreat inland. Some invest heavily in engineering. Some accept lower productivity in exchange for stability. Others fracture as regional differences widen and shared expectations fail.

The ocean remains alive.

It remains powerful.

But it becomes less collaborative.

This is the quiet consequence of weakening tides: not dramatic loss, but the slow erosion of synchrony—between organisms, between ecosystems, and between human systems and the natural cycles they once relied on without ever needing to acknowledge.

Extinction by reconfiguration is not loud.

It is a series of small mismatches that never quite resolve.

And by the time the pattern is obvious, it has already been in motion for longer than anyone can remember deciding to accept it.

That is how the long view works.

Nothing announces itself.

Everything adjusts.

And the cost is paid not in endings, but in the gradual disappearance of the conditions that once made continuity effortless.

∝ ∝ ∝

What Happens To Time Itself (Quietly Devastating)

The final change does not happen in the sky or the sea.

It happens in the clock.

Earth's rotation has never been constant. It has always been slowing, almost imperceptibly, as angular momentum leaks outward into the Moon's orbit without pause. The effect is so small it feels like trivia when expressed in numbers—milliseconds per century, fractions of seconds spread across generations.

But time is not experienced in fractions.

It is experienced in structure.

As Earth's rotation slows, days lengthen. Not suddenly. Not enough to notice within a lifetime. But enough that, over geological time, the length of a day becomes a variable rather than an assumption.

This has already happened before.

Hundreds of millions of years ago, days were shorter. Years contained more of them. Life adapted without awareness, because biology does not care about clocks. It entrains itself to light, to cycles, to whatever rhythm repeats reliably enough to become background.

Biological time stretches easily.

Circadian systems adjust. Metabolic cycles rescale. Seasonal cues shift with the Sun's changing arc. Organisms don't require a fixed number of hours. They require consistency.

Human systems are different.

Civilization depends not just on rhythm, but on synchronization across space and scale, people and institutions. On the assumption that a day is a day everywhere, that calendars align, that time can be shared, recorded, inherited, and enforced.

As Earth's rotation slows, that assumption becomes harder to maintain.

At first, the difference is absorbed quietly. Corrections are added. Adjustments are debated. Systems compensate without acknowledging that compensation is occurring. Timekeeping bodies argue about whether to insert, remove, or redefine units to keep atomic precision aligned with planetary motion.

The language is technical. The debates are procedural.

But the meaning underneath is simple.

Natural time and artificial time are drifting apart.

Atomic clocks don't slow. They tick with extraordinary regularity, indifferent to tides, oceans, or the Moon. Earth, by contrast, is a physical object responding to forces it cannot escape. Its rotation is messy.

influenced by mass redistribution, atmosphere, oceans, and the same lunar interactions that shape tides and axial stability.

The result is that "the day" stops being a clean inheritance from nature.

It becomes a negotiated construct.

Negotiation implies authority.

This negotiation begins as inconvenience. Systems that rely on synchronization—navigation, communication, financial markets—demand precision that Earth no longer supplies for free. Corrections multiply. Standards bodies intervene. Time zones remain, but their relationship to the Sun grows increasingly abstract.

Eventually, a decision is made. Coordination has become too costly.

Human timekeeping decouples.

The clock becomes something we impose.

This is quietly devastating because it severs a deep continuity. For most of human history, time was something you lived inside. Days followed nights. Seasons followed days. Calendars approximated the sky closely enough that no one worried about the difference.

As rotation slows and corrections accumulate, that inheritance fades.

Time becomes administrative.

Biological rhythms continue to follow light and darkness. Agriculture still depends on seasons. Bodies entrain themselves to cycles that remain planetary, not atomic. But the official time—the one used to coordinate societies—drifts further from those cues.

A split emerges.

Natural time persists.

Civil time dominates.

The two coexist, but they are no longer the same thing.

This does not cause collapse.

It causes friction.

People experience fatigue without clear cause. Systems require constant resynchronization. The effort required to keep everything aligned grows steadily, invisibly, year by year.

Again, nothing dramatic happens.

There is no day when time breaks.

There is only a long transition from time as something received to time as something maintained.

From the long view, this is the final withdrawal of subsidy.

The planet no longer provides a perfectly reliable metronome. It still turns. It still cycles. But not with the precision modern civilization demands.

So civilization compensates.

It builds clocks that ignore the sky. Calendars that prioritize coordination over alignment. Time standards that favor stability of systems over fidelity to Earth.

Time becomes a human artifact.

Accurate. Precise. Enforced.

And slightly unmoored.

This is the quiet culmination of everything that came before. The Moon's retreat, the loosening axis, the weakening tides—all of it feeds into this moment where the most basic structure of shared experience is no longer inherited from the world, but imposed upon it.

The planet continued doing exactly what it always has, long enough for the difference to matter.

From the long view, nothing here is catastrophic.

It is simply profound.

A civilization that must define its own time has crossed a threshold—not into collapse, but into a deeper awareness that the world it inhabits is not fixed, and never promised to be.

Time does not end.

It becomes managed.

And that, quietly, changes everything.

∝ ∝ ∝

What changes next is not measurement.

It is meaning.

Once time becomes something that must be maintained rather than inherited, it stops being neutral. It acquires authorship. Someone decides what counts as *now*. Someone decides which reference frame matters more: the sky or the system.

Those decisions are technical at first. Then they are institutional. Eventually, they are cultural.

For most of history, time was external. It arrived whether you believed in it or not. The Sun rose. The day began. The calendar followed the seasons closely enough that disagreement was rare and irrelevant.

When civil time detaches from planetary time, disagreement becomes possible.

Not about seconds or hours, but about authority.

Which clock governs labor?
Which time defines contracts?
Which reference frame determines responsibility?

These questions don't appear as philosophical debates. They appear as legal edge cases. As disputes over deadlines that no longer align cleanly with daylight. As systems that insist on precision even when human bodies resist it.

Biology continues to entrain itself to light and dark. Sleep, hormone cycles, attention—all remain stubbornly planetary. Civilization, meanwhile, increasingly asks bodies to comply with clocks that no longer track the sky.

This mismatch does not break people.

It taxes them.

Fatigue becomes ambient. Synchronization becomes effortful. The sense of shared rhythm—of everyone moving through the same day together—thins.

Again, no catastrophe.

Just erosion.

The deeper implication is not that time becomes false, but that it becomes *conventional.* Something agreed upon rather than given. That agreement holds only as long as coordination matters more than alignment with nature.

From the long view, this is another quiet renegotiation.

Civilization has already done this with distance, with borders, with value. Time is simply the last domain where inheritance gives way to enforcement.

And once time is no longer something received from the world, it changes how the world is experienced. Days feel longer or shorter because attention is strained. Seasons feel out of phase because calendars have stopped caring whether they match.

Nothing about this is wrong.

It is adaptive.

But it carries a cost that never appears on a ledger: the loss of a shared, external rhythm that once required no explanation.

From the long view, this is the quietest consequence of all.

Not that time runs out.

But that time, once universal and unquestioned, becomes one more system civilization must actively hold together—forever.

That is quietly devastating because it marks the moment when even the most basic structure of experience is no longer something the planet provides for free.

At that point, civilization is not just living *on* Earth.

It is actively managing its relationship to it.

WHAT DOES *NOT* HAPPEN

The final clarification matters because imagination fills gaps faster than physics ever could.

None of this ends in spectacle.

Earth does not crack or tear itself apart. There is no moment when continents fly loose or oceans spill into space. The Moon does not suddenly slip its bonds and vanish into darkness, leaving the planet shocked and broken. Gravity does not weaken, flicker, or fail. The rules remain intact. They are not exhausted. They don't forget how to work.

Life does not end.

That expectation—the suddenness, the violence, the finality—is a reflex born of stories that prefer collapse to continuity. But the long view offers something more restrained and more difficult to sit with.

Everything keeps functioning.

The planet continues to orbit the Sun. The Moon continues to orbit Earth. Gravity continues to bind mass to mass with the same indifference it always has. Biology continues to adapt, diversify, and persist. Ecosystems rearrange themselves again and again, not toward ruin, but toward whatever configurations the new conditions allow.

There is no failure mode hidden here.

There is only drift.

Drift is change without drama.

What changes is not the existence of the system, but the shape of the space it occupies. The conditions that once clustered tightly enough to feel permanent begin to spread out. The familiar corridors of stability widen. Patterns that once repeated reliably grow harder to inherit unchanged.

Nothing breaks, but nothing stays guaranteed.

This distinction is essential, because destruction implies error. It suggests that something went wrong, that a wrong choice was made, that intervention could restore an earlier state. What happens here does not belong to that category.

It is not damage. It is departure.

A departure from the conditions under which complex life—and later civilization—evolved and learned to assume continuity. Those conditions were never promised. They were never fixed. They were simply present long enough to be mistaken for permanence.

From the long view, the Earth–Moon system does not collapse.

It moves on.

The planet remains alive. It remains dynamic. It remains capable of supporting life in abundance. What it no longer does, over deep time, is hold still long enough for inherited assumptions to remain effortless.

Civilization struggles because the world becomes less familiar or habitual. Less willing to repeat itself. Less inclined to subsidize long-term planning with predictable rhythms. Institutions built on expectation find themselves negotiating variance instead. Skills optimized for stability lose efficiency when patterns drift.

What once felt like foresight begins to resemble guesswork.

This is not an ending.

It is a transition out of a narrow window of convenience and into a broader range of possibility—some benign, some demanding, none catastrophic in the way humans expect catastrophe to look.

The danger was never destruction.

The danger was mistaking stability for permanence.

Once that mistake is corrected, the long view becomes calm again—not reassuring, not threatening, just clear.

The world continues.

It simply continues without promising to meet us under the same conditions forever.

∝ ∝ ∝

The Philosophical Endpoint

This is where the long view stops being technical and becomes unmistakably philosophical—it follows the physics all the way out.

If the process is extended far enough—tens of millions of years, then hundreds—Earth does not become barren. It does not become silent. It does not become hostile to life.

It becomes unfamiliar to intelligence.

Life continues. It always does. It adapts to shifting light, migrating climates, changing chemistry. It fills niches as they appear and abandons them when they vanish. Biology has no expectation of fairness or repetition. It only requires opportunity.

What disappears, slowly and without drama, is consistency.

Seasons no longer arrive with the same confidence. Climate cycles lose their clean periodicity. The long, patient rhythms that once allowed prediction to feel like inheritance become irregular enough that memory stops conferring advantage.

Nothing forbids intelligence from arising in such a world.

But intelligence shaped by that world wouldn't resemble ours.

Human intelligence—our kind of intelligence—emerged inside repetition. Inside a planet that leaned the same way long enough for seasons to teach anticipation. Inside climates stable enough that agriculture could assume return. Inside tidal and daily rhythms steady enough to turn time into expectation rather than surprise.

The Moon didn't create life.

It created consistency.

It damped extremes. It smoothed oscillations. It narrowed the range of possibilities just enough that complex systems could build on top of one another instead of constantly starting over. It allowed the future to look enough like the past that planning became meaningful.

As that influence fades, Earth does not lose life.

It loses legibility.

The planet still receives sunlight. Still circulates water. Still hosts ecosystems of staggering complexity. But the patterns no longer repeat cleanly enough to support long chains of expectation. Intelligence, if it persists, adapts toward immediacy. Toward flexibility over memory. Toward response rather than foresight.

Civilization, as we understand it, becomes harder to sustain because continuity vanishes.

This is the endpoint that rarely appears in stories because it lacks spectacle. There is no collapse to film. No disaster to count down. Only a gradual transition from a world that quietly rewarded planning to one that quietly refuses to.

Earth remains alive.

It remains beautiful.

But it becomes a planet where stability is local, temporary, and negotiated—never assumed.

From the long view, this is not tragedy.

It is consequence.

The Moon's gift was never life itself. Life is stubborn enough to find a way almost anywhere. The gift was the narrowing of variance—the creation of a background rhythm steady enough for intelligence to mistake it for permanence.

As that rhythm stretches and softens, intelligence does not vanish.

It changes.

And the deepest realization—the one this block leaves the reader with—is not about loss, but about gratitude delayed too long:

That the quiet regularity we built everything on was never guaranteed.
That it was borrowed from a configuration that happened to hold.
And that the future does not owe us the same stillness we were given.

That is the philosophical endpoint.

Not extinction.

But the end of the assumption that the world will always meet us halfway.

∝ ∝ ∝

Why This Terrifies Institutions (And Should)

This is where the calm becomes dangerous.

The implications change.

Institutions are built on horizons. Not just budgets and election cycles, but assumptions about continuity: that the ground beneath long-term planning will remain sufficiently familiar, sufficiently stable, that deferral is rational. Problems can be postponed. Trade-offs can be managed later. Time itself is treated as an ally.

The long view removes that ally.

What terrifies institutions is not the prospect of disaster. They are designed to survive disasters. They have playbooks for shocks, collapses, emergencies, and recovery. Those scenarios preserve a comforting logic: something breaks, something responds, something is restored.

This offers none of that.

It offers no rupture to organize around.
No moment to declare response.
Only continuation that erodes certainty while appearing benign.

Earth is not permanently stable.
It is temporarily fortunate.

That distinction is corrosive to systems built on inevitability narratives—nations, markets, ideologies that quietly assume the background conditions of human success are fixed features of reality rather than historical coincidences.

Exceptionalism relies on permanence.
Permanence relies on repetition.
Repetition relies on physics continuing to cooperate.

Here, physics does not rebel. It simply declines to guarantee.

There is no villain to confront. No policy lever to pull. No innovation that reverses angular momentum transfer or re-tightens axial damping. There is no "later" that fixes orbital mechanics.

That is the part institutions cannot metabolize.

Because once the background is revealed as contingent, delay stops being neutral. "We'll figure it out later" loses its comfort when later does not restore earlier conditions—only more drift, more variance, more negotiation with a world that no longer repeats itself cleanly.

This does not demand immediate action.

It demands acknowledgment.

And acknowledgment is destabilizing.

It undermines the quiet story every large system tells itself: that history bends toward improvement because the stage remains constant. That progress is additive because the rules of the game don't change beneath it. That tomorrow will look enough like today for today's promises to make sense.

The long view says otherwise.

It says the stage moves.

Slowly.

Relentlessly.

Without consultation.

Institutions fear this because it dissolves the moral cover of deferral. If the world is not fixed, then postponement is not prudence—it is choice. And choice implies responsibility.

That is why this terrifies them.

Because it removes the illusion that stability is owed.

A temporarily fortunate planet offers no guarantees beyond the moment it is observed. And systems that depend on guarantees—of climate, of time, of repetition—must either adapt their philosophy or cling harder to narratives that pretend the future will wait.

The fear is not of extinction.

It is of honesty.

Because once it is admitted that the world we built on was a gift rather than a constant, institutions are forced to confront a question they are not structured to answer:

What does stewardship mean when the foundation itself is moving—and there is no version of "later" that makes it stop?

PHASE 7 — LIVING INSIDE THE LONG NOW

No one marked the moment when the long view became ordinary.

By then, the absence of resolution was no longer a failure of action, but a condition that had been accepted.

There was no announcement, no press conference where someone finally said *this is how things are now*. The world didn't pause to absorb the implication. People still woke up, still checked the weather, still made plans that assumed tomorrow would resemble yesterday closely enough to be useful.

And for the most part, it did.

That was the strange thing. Nothing in daily life contradicted what had been learned. The Sun still rose. Tides still moved. Seasons still arrived. The planet continued to feel familiar enough that urgency never quite found a foothold.

What changed wasn't behavior at first.

It was orientation.

Scientists stopped framing their work as answers and began framing it as boundaries. Papers grew more careful, not in language but in ambition. Models were no longer presented as forecasts, but as envelopes—spaces within which futures might unfold rather than paths they would follow.

Planners learned a new word without ever formalizing it: *variance*.

Not as a parameter to be minimized, but as a condition to be endured. Something you didn't eliminate, only widened margins around and learned to live inside.

Infrastructure was no longer designed around optimal conditions, but around ranges. Margins widened. Flexibility replaced efficiency in places where efficiency had once been sacred. Systems were built with exit ramps instead of endpoints, designed to be revised rather than completed.

Nothing about this felt heroic.

It felt administrative.

Ethics followed, slowly and without consensus.

The old moral frameworks had assumed a stable stage. Protect the future. Preserve what exists. Leave things better than you found them. Those imperatives depended on an unspoken promise: that the future would be structurally similar enough to receive what was passed down.

Once that promise dissolved, responsibility became harder to define.

Was it ethical to preserve systems that required predictability the planet no longer offered for free? Was it ethical to invest in permanence knowing it would demand increasing intervention from generations who had not agreed to pay that cost? Was honesty an obligation if it changed nothing—or an obligation *because* it changed nothing?

There were no clean answers. Only competing goods.

Some argued that stewardship now meant adaptation above all else: teaching societies how to move, reorganize, abandon, and rebuild without trauma. Others argued that preserving continuity, even if artificially, was itself an ethical act—that shielding people from variance was no different from shielding them from disease or famine.

Both positions were rational.

Neither resolved the tension.

Civilization absorbed this the way it absorbs everything else: unevenly.

Some cultures leaned into flexibility. They built lightly, planned provisionally, treated institutions as living systems meant to be revised rather than defended. Others doubled down on control, investing heavily in engineering, synchronization, and artificial stability to preserve inherited rhythms as long as possible.

Neither path failed outright.

Neither path solved anything permanently.

Over time, the idea of permanence itself began to feel dated—not wrong, but provincial. A relic of a period when the background held still long enough for monuments to make sense. Younger generations didn't reject continuity; they redefined it. Continuity was no longer about sameness. It was about coherence across change.

To remain intelligible to your descendants mattered more than remaining unchanged.

That shift was quiet, but profound.

Civilization didn't end. It thinned, diversified, regionalized. Long chains shortened. Global assumptions gave way to local negotiations. Memory became something curated rather than assumed, taught deliberately rather than inherited passively.

Time was still measured. Seasons were still named. But everyone understood, at some level, that these were conveniences—useful abstractions imposed on a planet that no longer promised to align with them forever.

And still, life went on.

Children learned history. Engineers built bridges. Farmers planted crops. The difference wasn't despair. It was humility. A recognition that the world had never been obligated to cooperate, and that cooperation, when it occurred, was something to notice rather than take for granted.

The Moon continued to move away.

Slowly enough that no one felt it. Surely enough that no one could deny it.

It didn't take life with it. It didn't take meaning.

For most of human history, the Moon appeared fixed. It rose. It set. It returned in cycles that could be named, charted, ritualized. It governed calendars, tides, and stories. Whatever changed did so slowly enough to pass as permanence.

When measurement became precise enough to detect motion, the discovery didn't arrive with alarm. The Moon wasn't unstable. Its orbit wasn't failing. Nothing had gone wrong. It was simply farther away than it had been before, by a distance small enough to be expressed in centimeters, and large enough to accumulate into certainty.

The finding was absorbed easily. It required no revision of daily life. No behavior depended on the correction. The models accounted for it. The data held. The Moon was moving away from Earth at a measurable, predictable rate.

What was less frequently asked was why.

There was no external force acting on the system. No hidden mass, no approaching body, no unseen pull. The mechanism was ordinary, well understood, and mathematically unremarkable. The Earth's rotation, interacting with the Moon's gravity, transferred energy outward. Momentum conserved itself. Distance increased. Time adjusted. The system behaved exactly as physics said it should.

Nothing was pulling the Moon away.

The Earth was doing it to itself.

This wasn't a flaw in the system. It was the system. A slow redistribution of energy, proceeding continuously, without threshold or inflection point. The process didn't accelerate toward crisis. It didn't reverse. It didn't require intervention. It simply continued, converting rotation into separation, day length into distance, presence into absence measured too gradually to feel.

The implications unfolded on timescales that resisted narrative. The lengthening of days. The reshaping of tides. The distant future where the Moon would appear smaller, dimmer, quieter. These were not problems to be solved. They were consequences to be lived inside of.

The knowledge didn't demand action because there was no moment when action would have mattered. By the time measurement revealed the change, the process was already ancient. By the time it was understood, it had already normalized itself into background truth.

Civilizations continued. Institutions recorded the data. The Moon kept moving.

What made this different from catastrophe wasn't its scale, but its patience. There was no failure state to point to, no warning signal to interpret, no emergency to convene around. The system didn't collapse.

It adapted. Language stabilized around the fact. Responsibility diffused across time. Attention shifted elsewhere.

The experiment continued.

This is how long systems change: not by breaking, but by obeying their own rules without interruption. Not by surprise, but by persistence. Not by external pressure, but by internal consistency carried forward long enough to escape notice.

The Moon does not announce its departure.
The Earth does not feel itself slowing.
The distance grows anyway.

And this is what it means to live inside a long view—not to witness an ending, but to recognize that continuity itself has consequences, even when nothing seems to be happening at all.

What it took was the illusion that the conditions which shaped intelligence were permanent features of reality rather than a long, generous coincidence.

Once that illusion faded, humanity didn't collapse.

It adjusted its posture.

Not toward fear.
Not toward resignation.
But toward a quieter understanding:

That living well was never about fixing the world.
It was about learning how to live inside motion without pretending it would stop.

This was the final inheritance of the long view.

Not a warning.
Not a prophecy.

Just the knowledge that the present was never a destination—only a momentary alignment in a system that would keep going, with or without an audience.

And that meaning, if it was going to persist at all, would have to be made flexible enough to move with it.

∝ ∝ ∝

The adjustment didn't feel like surrender.

It felt like learning a different posture—one that didn't brace against change or rush to outrun it, but learned how to stand inside it without losing balance.

Institutions didn't disappear. They changed their tone. Fewer proclamations. More revisions. Fewer claims of permanence. More acknowledgments of limits. Documents that once ended in certainty began ending in review dates and contingencies that were not treated as failure, but as honesty.

The most resilient systems were not the most ambitious ones.
They were the ones that expected to be revisited—designed with revision, to be adjusted rather than refined.
Their strength came not from reaching an endpoint, but from assuming none would ever hold.

Education shifted almost without comment. History was taught less as a sequence of achievements and more as a sequence of adaptations. Students learned not just what had worked, but *why it stopped working.* Collapse was no longer framed as moral failure, nor success as inevitability. Both were treated as responses to conditions that didn't stay still.

This didn't produce cynicism.

It produced restraint.

The idea of progress didn't vanish, but it narrowed. Progress was no longer measured by expansion alone, but by how gracefully systems could change shape without harming the people inside them. A good institution wasn't one that lasted unchanged, but one that could be dismantled without catastrophe.

Ethics matured under this pressure.

Responsibility stopped meaning "make it permanent" and began meaning "don't leave behind something that can only survive if nothing changes." It marked a shift in commitment—not in consequence, but in how the future was allowed to arrive. It became a question of what could adapt, what could be revised, and what could be allowed to evolve without breaking the structures built around it. Stewardship became less about endurance over time, rigid continuity, or preservation and more about easing transition—about leaving instructions instead of monuments.

This was hardest for those who remembered certainty.

Easiest for those who grew up without expecting it.

You could see the difference in how people spoke about the future. Fewer grand arcs. More near horizons stitched together carefully. The future was no longer a destination to be reached, but a relationship to be maintained.

Even language softened around time.

Deadlines were still enforced, but fewer people pretended they were natural. Calendars were still followed, but no one confused them with the sky any longer. Time was understood as a tool—necessary, imperfect, and always in need of calibration.

This didn't cheapen meaning.

It clarified it.

When the world is not assumed to be stable, attention sharpens. Care becomes deliberate. Gratitude becomes less abstract. The present is no longer treated as a holding pattern before something better arrives. It is recognized as a brief coherence worth inhabiting fully. Presence becomes an ethical act rather than a sentimental one.

The Moon kept moving.

That fact stopped being unsettling and became background again—not because it no longer mattered, but because it had been integrated. The long view was no longer a revelation. It was context.

Humanity didn't become wiser in any absolute sense. It remained contradictory, uneven, capable of cruelty and care in equal measure. What changed was the scale at which wisdom was expected.

Fewer people believed in final answers. More people believed in continuity through change. Understanding shifted from being something to arrive at, to something that had to be maintained.

Civilization, in this form, was lighter.

Less certain.

More adaptable.

Still flawed.

But no longer pretending that the ground beneath it was fixed.

If there was an ending here, it wasn't closure.

It was orientation.

A collective, unspoken agreement that the task wasn't to arrest motion, or to wait for a moment when everything would finally settle, but to live well inside a world that never promised to.

The long now didn't demand heroism.

It demanded attention.

And that, in the end, was something humanity had always been capable of—once it stopped expecting the universe to hold still long enough to make the effort optional.

AUTHOR'S NOTE

This book is not speculative fiction.

Everything described here rests on established, publicly available science: measurements of the Earth–Moon system, well-understood principles of orbital mechanics, long-studied interactions between axial tilt and climate, documented changes in Earth's rotation, and widely accepted models of deep-time planetary behavior. No new physics is proposed. No catastrophic predictions are made.

What this book attempts is synthesis.

Modern science is extraordinarily good at precision and specialization. It measures, models, and publishes with rigor. What it does less often—by necessity rather than neglect—is assemble slow, distributed truths into a single narrative that can be held in the mind without simplification or drama.

The phenomena discussed here unfold across timescales that don't align neatly with politics, markets, or daily experience. Because of that, they are often treated as background conditions: accurate, important, and safely deferred. This book does not challenge their validity or exaggerate their effects. It simply refuses to treat them as conceptually separate.

Nothing in these pages argues that the Earth is ending, that catastrophe is imminent, or that humanity faces an unavoidable collapse. Life on this planet is resilient. Physics is stable. The future described here is long, not cinematic.

What *is* examined is a quieter implication: that some of the conditions under which human civilization evolved—predictable seasons, stable axial behavior, inherited rhythms of time—are not permanent features of reality, but outcomes of a particular planetary configuration that has been slowly changing for billions of years.

This is not an argument for panic or for policy. It is an argument for orientation.

The book is written as a narrative because narrative is how humans integrate meaning across scale. The prose is restrained by design. There are no villains, no secret knowledge, and no call to action disguised as revelation. Institutions are portrayed not as deceptive, but as rational actors operating within incompatible horizons of responsibility.

If the book unsettles, it is because it removes an assumption many of us carry without realizing it: that the background of human success is fixed.

It is not.

It never was.

This book does not ask the reader to fear that fact. It asks only that it be acknowledged.

What follows in the references section documents the scientific foundations of this work. What remains after reading is not an answer, but a posture—a way of standing inside a world that moves slowly, continuously, and without regard for our expectations.

That posture, more than any conclusion, is what this book is about.

—Antonio Pascarella

On Scope and Scale

This book operates across scales that don't sit comfortably together.

It moves from decades of measurement to millions of years of consequence, from institutional decision-making to planetary mechanics, from lived human time to deep geological time. That movement is intentional, and it requires clarification.

The science referenced here spans multiple domains: orbital dynamics, geophysics, climate science, oceanography, and timekeeping. Each of these fields is internally rigorous and externally cautious. None of them, on their own, claim to describe the future of civilization or the fate of intelligence. This book does not ask them to.

Instead, it places their conclusions side by side.

No single discipline is extended beyond its accepted limits. Where quantitative certainty exists, it is respected. Where uncertainty is structural—arising from chaotic systems, long timescales, or coupled interactions—it is treated as a feature rather than a flaw.

For this reason, the book avoids precise long-term numerical forecasts. Over spans of millions of years, specificity implies confidence that science itself does not claim. The intent here is not to predict exact outcomes, but to describe the *directional behavior* of well-understood systems when allowed to operate over sufficient time. What mattered wasn't making the moment last, but noticing it while it held.

Timescales shift deliberately throughout the text.

Short horizons are used where human institutions, politics, and planning operate. Long horizons are used where physical processes dominate and human categories lose their relevance. The transitions between these scales are not smoothed over, because the friction between them is part of the subject itself.

Uncertainty is not used as a rhetorical shield.

Where the book describes future conditions, it does so conservatively and qualitatively, grounded in established mechanisms rather than speculative scenarios. The absence of dramatic timelines or thresholds is not an omission; it reflects the nature of the systems involved. Many of the most consequential changes discussed here don't manifest as events, but as gradual losses of predictability.

This book also makes a deliberate distinction between *life* and *civilization.*

Biological resilience and civilizational continuity operate under different constraints. Life adapts across variability; civilizations depend on repeatability, inheritance, and long-term coordination. The book does not conflate these, and it does not treat human structures as synonymous with planetary habitability.

Finally, this is not a policy document.

No prescriptions are offered, and none are implied. The book does not argue for specific actions, technologies, or governance frameworks. Its purpose is orientation, not instruction—to assemble slow truths into a coherent frame that can be acknowledged without exaggeration or denial.

The methodology, in short, is conservative in science and expansive in synthesis.

Everything here is already known, somewhere.

The work of this book is to hold those pieces together long enough for their collective implication to be felt.

References

The scientific material discussed in this book is drawn from long-standing, peer-reviewed research programs and publicly available institutional sources. The references below represent foundational work and representative entry points rather than an exhaustive bibliography.

Lunar Laser Ranging and Earth–Moon Distance

- Jet Propulsion Laboratory (JPL), NASA
  Lunar Laser Ranging Experiment
  https://science.jpl.nasa.gov/projects/lunar-laser-ranging/
- Williams, J. G., Turyshev, S. G., & Boggs, D. H.
  "Lunar Laser Ranging Tests of the Equivalence Principle."
  Classical and Quantum Gravity, 2009.
- Apache Point Observatory Lunar Laser-ranging Operation (APOLLO)
  https://www.apollo-observatory.org/
- Dickey, J. O. et al.
  "Lunar Laser Ranging: A Continuing Legacy of the Apollo Program."
  Science, 1994.

Orbital Mechanics and Earth–Moon Dynamics

- Murray, C. D., & Dermott, S. F.
  Solar System Dynamics.
  Cambridge University Press, 1999.
- Lambeck, K.
  The Earth's Variable Rotation: Geophysical Causes and Consequences.
  Cambridge University Press, 1980.
- Touma, J., & Wisdom, J.
  "Evolution of the Earth–Moon System."
  The Astronomical Journal, 1994.

Axial Tilt, Obliquity, and Long-Term Climate Behavior

- Laskar, J., Robutel, P., Joutel, F., et al.
  "A Long-Term Numerical Solution for the Insolation Quantities of the Earth."
  Astronomy & Astrophysics, 2004.
- Laskar, J., & Robutel, P.
  "Stability of the Planetary Three-Body Problem."
  Nature, 1993.
- Ward, W. R.
  "Obliquity Variations of Mars."
  Science, 1974.

Tides, Oceans, and Coastal Mixing

- Wunsch, C., & Ferrari, R.
  "Vertical Mixing, Energy, and the General Circulation of the Oceans."
  Annual Review of Fluid Mechanics, 2004.
- Munk, W., & Wunsch, C.
  "Abyssal Recipes II: Energetics of Tidal and Wind Mixing."
  Deep-Sea Research, 1998.
- Pugh, D. T., & Woodworth, P. L.
  Sea-Level Science: Understanding Tides, Surges, Tsunamis and Mean Sea-Level Changes.
  Cambridge University Press, 2014.

Earth's Rotation and Timekeeping

- International Earth Rotation and Reference Systems Service (IERS)
  https://www.iers.org/
- McCarthy, D. D., & Seidelmann, P. K.
  Time: From Earth Rotation to Atomic Physics.
  Wiley-VCH, 2009.
- Nelson, R. A., McCarthy, D. D., et al.
  "The Leap Second: Its History and Possible Future."
  Metrologia, 2001.
- NASA Earth Observatory
  Earth's Rotation and Length of Day
  https://earthobservatory.nasa.gov/

Deep Time, Habitability, and Planetary Context

- Kasting, J. F., Whitmire, D. P., & Reynolds, R. T.
  "Habitable Zones Around Main Sequence Stars."
  Icarus, 1993.
- Walker, J. C. G., Hays, P. B., & Kasting, J. F.
  "A Negative Feedback Mechanism for the Long-Term Stabilization of Earth's Surface Temperature."
  Journal of Geophysical Research, 1981.

Notes on Use

The works listed above inform the physical mechanisms, measurements, and constraints discussed throughout the book. Interpretations regarding long-term implications, civilizational dependence on predictability, and philosophical framing are the author's synthesis, grounded in these established findings.

Image Credit

Cover image:

ACKNOWLEDGMENTS

This book draws on the work of thousands of scientists, engineers, and observers whose efforts span decades and, in some cases, entire careers. Many of them will never read this book. Most of them didn't set out to create a narrative. They set out to measure carefully, publish honestly, and allow results to stand even when they didn't invite urgency.

That discipline made this book possible.

I am grateful to the researchers and institutions who have maintained long-running measurement programs—often quietly, often without public attention—because continuity, not novelty, is what allows slow truths to become visible at all.

Thanks are also due to educators and communicators who have preserved accuracy without exaggeration, and to those who resisted the temptation to simplify complex systems into spectacle. Their restraint shaped the tone of this work more than any stylistic choice.

I acknowledge the influence of colleagues, reviewers, and early readers who challenged clarity where it was assumed and pressed for precision where language threatened to soften implication. Any remaining imprecision is my responsibility alone.

Finally, I thank those who understand that not every important story arrives as an event, and that some of the most consequential work of science is done in patience rather than urgency.

About the Author

Antonio Pascarella is a systems-oriented writer and observer whose work focuses on the long-term interaction between physical reality, human institutions, and inherited assumptions of stability.

His background spans technology, organizational systems, and applied analysis, with a long-standing interest in how slow, cumulative processes shape outcomes that appear sudden only when viewed too narrowly. He has worked across domains where precision, continuity, and long-horizon thinking are required, and where uncertainty must be managed rather than eliminated.

This book grew out of sustained engagement with publicly available scientific research, particularly long-running measurement programs whose significance emerges only when considered collectively and over time. Pascarella's approach emphasizes synthesis over speculation and narrative clarity over dramatization.

He also enjoys astrophotography, and the lunar image used on the cover was captured by him during independent observation and has been previously published.

He lives and works in the United States.

www.ingramcontent.com/pod-product-compliance
Lightning Source LLC
LaVergne TN
LVHW020415220926
848484LV00019B/2582/J